Effect of IL-10 and anti-TGF-beta antibodies on the morphology of bone marrow stroma cultures
from
Interleukin-10
by
Jan E. DeVries and René de Waal Malefyt

Molecular Biology Intelligence Unit

"Glucokinase": its Regulation and Role in Liver Metabolism

MOLECULAR
BIOLOGY
INTELLIGENCE
UNIT

"GLUCOKINASE": ITS REGULATION AND ROLE IN LIVER METABOLISM

María Luz Cárdenas, Ph.D.

Laboratoire de Chimie Bactérienne
Marseille, France

R.G. LANDES COMPANY
AUSTIN

Molecular Biology Intelligence Unit

"GLUCOKINASE": ITS REGULATION AND ROLE IN LIVER METABOLISM

R.G. LANDES COMPANY
Austin, Texas, U.S.A.

Submitted: February 1995
Published: June 1995

Please address all inquiries to the Publisher:
R.G. Landes Company, 909 Pine Street, Georgetown, Texas, U.S.A. 78626
or
P.O. Box 4858, Austin, Texas, U.S.A. 78765
Phone: 512/ 863 7762; FAX: 512/ 863 0081

U.S. and Canada ISBN 1-57059-207-1

International ISBN 3-540-59285-7

Library of Congress Cataloging-in-Publication Data

Cárdenas, María Luz.
Glucokinase : its regulation and role in liver metabolism / by María Cárdenas.
p. cm. — (Molecular biology intelligence unit)
Includes bibliographical references and index.
ISBN 1-57059-207-1 (alk. paper)
1. Glucokinase—Regulation. 2. Liver—Metabolism—Regulation.
I. Title. II. Series.

QP606.G57C37 1995 95-18869
612.3'5—dc20 CIP

Publisher's Note

R.G. Landes Company publishes five book series: *Medical Intelligence Unit, Molecular Biology Intelligence Unit, Neuroscience Intelligence Unit, Tissue Engineering Intelligence Unit* and *Biotechnology Intelligence Unit.* The authors of our books are acknowledged leaders in their fields and the topics are unique. Almost without exception, no other similar books exist on these topics.

Our goal is to publish books in important and rapidly changing areas of medicine for sophisticated researchers and clinicians. To achieve this goal, we have accelerated our publishing program to conform to the fast pace in which information grows in biomedical science. Most of our books are published within 90 to 120 days of receipt of the manuscript. We would like to thank our readers for their continuing interest and welcome any comments or suggestions they may have for future books.

Deborah Muir Molsberry
Publications Director
R.G. Landes Company

Dedication

To Athel, and to the memory of Hermann Niemeyer.

CONTENTS

FOREWORD

This book describes an enzyme that has long interested both physiologically and kinetically minded biochemists, the former because of its importance in liver function, and the latter because of its unusual kinetics for a monomeric enzyme. It provides an example of a single gene that codes for two different proteins, producing a pancreatic enzyme that is slightly different from the one that occurs in the liver, and during the last few years the development of molecular genetics has given a new lease of life to the subject.

Hexokinase D (EC 2.7.1.1, often called "glucokinase") is the characteristic isoenzyme of hexokinase in the liver, and it is responsible for glucose phosphorylation in hepatocytes. Its importance derives from the fact that glucose uptake by the liver is an essential physiological process, crucial for glucose homeostasis. This point will be present throughout the book.

Hexokinase D has attracted attention since its discovery because of the effects of diet and hormones on its level of activity. These effects correspond to changes in the amount of enzyme, i.e. in gene expression, and will be described in Chapter 1 and further developed in Chapter 8, where the molecular bases for gene expression under different conditions will be discussed.

The name "glucokinase," together with the earlier classification as EC 2.7.1.2, contrasts with treatment of the other isoenzymes in higher organisms, which have always been known under the broader denomination of "hexokinases" (EC 2.7.1.1). The implication that it is an enzyme specific for glucose, like certain bacterial glucokinases, is quite misleading, as vertebrate "glucokinase" can also phosphorylate other hexoses, such as mannose, 2-deoxyglucose and fructose. In Chapter 2, I shall describe the origin of this misleading name and discuss why it is misleading.

The fact that the name glucokinase is misleading has created the difficult problem of deciding what to call the enzyme in the book. In the title, and in the section titles of the first chapter. I have stayed with "glucokinase," albeit with quotation marks, because this is the name that most biochemists use, whether rightly or wrongly, and I want potential readers to know what the book is about. Within the book itself, using quotation marks throughout would certainly be so irritating to many readers that it would be self-defeating. Omitting the quotation marks, on the other hand, would perpetuate an error and lead to confusion in the contexts where I need to refer to the genuine glucokinases that are found in bacteria and other lower organisms. I have opted, therefore, for the name that is most satisfactory in the research literature and is the least objectionable, namely hexokinase D. Provided that the reader realizes that this name refers to the enzyme that many people call glucokinase (or sometimes hexokinase IV) there should be no misunderstanding. In other publications, where the name glucokinase remains very common, readers should try to read it with quotation marks even if they are not written, unless, of course, the article is about an enzyme that is really specific for glucose.

As regulation of an enzyme by changes in its total activity is a long-term form of regulation, one might expect that additional and faster mechanisms should exist to allow fine adjustment of the activity of hexokinase D in response to the physiological conditions of variable input of glucose. As will be described in Chapter 3, the kinetic characteristics of hexokinase D allow this fine adjustment. Hexokinase D is half-saturated at a high concentration of glucose that is within the range of physiological concentrations of glucose arriving in the liver; moreover, the effect of this high value is supplemented by kinetic cooperativity with respect to glucose. This kinetic cooperativity is particularly interesting, as discussed in Chapter 6, not only because of its physiological role of increasing the sensitivity of the response to varying glucose concentrations, but also because hexokinase D is a monomeric enzyme with a single binding site for glucose, which means that the traditional binding models to explain cooperativity cannot be applied.

Hexokinase D accounts for more than 85% of the total glucose phosphorylating activity of the liver, although, as described in Chapter 1, it is just one of four hexokinase isoenzymes in the liver. Moreover, as presented in Chapter 4, all four of them are present in the hepatocyte, the only type of liver cell in which the hexokinase D gene is expressed. It follows that hexokinase D gene expression is specific for one cell type and active hexokinase D has been unambiguously detected only in hepatocytes and pancreatic β-cells. Despite this, several authors have described a glucose phosphorylating activity of low affinity in various tissues that they have identified as "glucokinase" (meaning hexokinase D), but in reality this enzyme is not a hexokinase D but an *N*-acetylglucosamine kinase. The reasons for the confusion and the precautions needed for avoiding it are described in Chapter 4.

In this book I deal mainly with liver hexokinase D. However, as hexokinase D plays an important role in controlling glucose phosphorylation and metabolism in both the liver and pancreatic islets, and as metabolism controls hexose-induced insulin release, the metabolic role of liver hexokinase D cannot be properly analyzed without reference to the pancreatic enzyme. Pancreatic hexokinase D is the major enzyme that phosphorylates glucose upon entry into islet β-cells. The slight differences in amino acid sequence from the liver enzyme that result from different processing of the same gene are described in Chapter 5. The important role played by hexokinase D in regulating insulin secretion and the uptake of glucose by the liver has led its gene to be identified as a prime candidate for one whose genetic variation or altered regulation could contribute to the development of type 2 diabetes, or non-insulin-dependent diabetes mellitus (NIDDM). Chapter

5 therefore also describes the connection between hexokinase D and NIDDM, which is responsible for much of the current interest in hexokinase D.

For a long time the control by substrate concentration represented the only known type of fast regulation of hexokinase D. It is now clearly established that there is an additional short-term regulation by the action of a regulatory protein; this will be presented in Chapter 7.

Some readers may feel that I pay too much attention in this book to the history of our knowledge and understanding of hexokinase D, and may feel that there is no reason for the modern experimenter to know why we had particular ideas—many now discarded—at particular times. However, I make no apology for this emphasis, because we can avoid repeating the mistakes of the past only if we understand why they were made in the first place. The possibility of large-scale production of mutant forms of hexokinase D with selected properties has opened the way toward many interesting experiments that could not be done in the 1970s; but if these are done without understanding of the behavior that led to confusion between hexokinase D and *N*-acetylglucosamine kinase, and of the precautions that need to be taken for ensuring that the cooperativity is not an artifact, the same confusion will certainly occur again.

I want to end by quoting some words of Dr. S. A. Kauffman from the Preface to his recent book The Origins of Order, which express my own thoughts very accurately: Authors know that books are not easily written. This one has been no exception. Yet writing a scientific book can be like writing a novel. Ideas, like characters, once loose upon a page harbor their own unsuspected paths, and mingle with their own logic. If useful, they have progeny.

María Luz Cárdenas

Acknowledgments

I am grateful to the Faculty of Sciences of the University of Chile for appointing me to the visiting Chair set up in memory of the late Professor Hermann Niemeyer Fernández. Professor Niemeyer was one of the pioneers in the study of hexokinase D, and introduced me to it when I was his student; his spirit is very much present in the book. Later visits to his former laboratory in Santiago have been very stimulating for writing the book, and I thank the members of the laboratory, especially Dr. Tito Ureta, for giving me the opportunity to do some of the work on it there, and for many useful discussions. I also thank Drs. Ken Neet (University of Chicago, USA), Louis Hue, Emile Van Schaftingen (both of the Catholic University of Louvain, Belgium) and Ian Trayer (University Birmingham, UK) for their prompt and helpful answers to my queries. I am likewise grateful to the R. G. Landes Company for encouraging me to write this book and for their patience in allowing me time to complete it.

The book has also greatly benefited from the criticisms of my colleague Dr. Athel Cornish-Bowden, and from his aid in preparing the figures and tables; his invaluable help and encouragement have made the whole project possible. (He is my husband as well as my colleague, and I am grateful to hexokinase D for bringing about our meeting). I would also like to thank my daughter Isadora for her patience and understanding during the many hours she has spend in the laboratory over the past months to allow this book to be completed, and for her constant love and support. Finally I thank Mr. J. Victor, the librarian of the CNRS campus in Marseilles, and his colleague Mrs. M. Charpin, for help in tracking down many of the references.

María Luz Cárdenas

CHAPTER 1

Diet and Carbohydrate Utilization: "Glucokinase," an Adaptive Enzyme

The general concept that the enzyme pattern in animal tissues can be altered by a variety of modifications in the environment of the animal is well established. The amount and quality of the food consumed is one of the important variables that may affect an organism, and in fact adaptation to different diets is an outstanding feature of most organisms that is directly related to their survival.

This chapter is a brief historical account of the original studies of glucose utilization that led to the discovery of hexokinase D "glucokinase" in liver and the establishment of its adaptive character. More details of the early work may be found in reviews by Niemeyer,[1,2] Walker[3] and Weinhouse,[4] of which reference 2 is particularly useful.

HISTORICAL BACKGROUND

More than a century ago, Claude Bernard was in 1859 the first to describe the influence of dietary conditions on carbohydrate utilization, after his discovery of liver glycogen and its role in maintaining the levels of glucose in the blood. He found that dogs fasted for 24 or 36 hours responded in a manner similar to diabetic animals when food was administered, i.e. with hyperglycemia and glycosuria, indicating impaired capacity to utilize glucose at a normal rate. In 1877 he generalized this observation, stating that "if a man or an animal is fasted for some time, and then given a good meal with an abundance of carbohydrate, glucose will appear in the urine," just as in diabetes. Further quantitative contributions were made in 1890 by Hofmeister, who coined the term "hunger diabetes." In the years that followed there were many studies in different species of animals, including man, with results that supported the idea that fasting and a decrease in the carbohydrate content of the diet reduce the capacity of the body to use glucose.

Understanding of the mechanism of hunger diabetes at the cellular level started with the introduction of isotopic compounds and of the technique of Warburg, which used tissue slices for studies of metabolism in vitro. Especially significant were the observations made about forty years ago in the laboratories of Chaikoff and Hastings. They incubated liver slices with [^{14}C]-glucose, taking these from animals subjected to different dietary and hormonal conditions, and they studied glucose utilization by analyzing some products of glucose metabolism. Their basic assumption in these experiments was that the prior treatment induced certain metabolic alterations that should modify the metabolic behavior of the isolated tissues. It was shown that liver slices from rats which had been deprived of carbohydrate for several days used glucose at a lower rate than liver slices from control animals. Thus there was a decrease in the oxidation to CO_2, as well as a diminution of glycogen and fatty acid synthesis. If the rats were given galactose or fructose instead of being completely deprived of carbohydrates, the liver slices also showed a diminution of the utilization of glucose. References to all these early studies may be found in the review by Niemeyer et al.[2]

These and other results indicated that under the conditions described there was an impairment in liver at an early step of glucose utilization, and the most plausible candidate for this step was the reaction responsible for the initial phosphorylation of glucose. Stimulated by these types of results, and by the idea that the enzyme composition of liver cells might vary under different circumstances such as changes in the diet, Hermann Niemeyer and his group in Chile decided to explore the effects of modifying the amount and nature of food on the levels of several enzymes directly involved in the utilization of glucose, and these proved to show widely different responses to dietary changes.[5] Some enzymes, like glycogen phosphorylase, glycogen synthetase and the glucose phosphorylating activity, depended strictly on the carbohydrate supply; others, such as glucose 6-phosphatase, depended on the amount of protein given; the activity of yet another group of enzymes, including glucose 6-phosphate dehydrogenase, appeared to depend on the simultaneous provision of both protein and carbohydrate. Finally, it seemed as if some enzymes, like glucose 6-phosphate isomerase, remained constant under all dietary conditions studied.

The glucose phosphotransferase activity in liver was highly dependent on dietary carbohydrate in different tests.[6,7] For example, maximal levels were observed when carbohydrate was present in the diet, and lowered levels when protein or fat was substituted for carbohydrate in isocaloric diets. There was a very good correlation between the specific activity of the enzyme (units per gram of liver protein) and the proportion of carbohydrate in the diet. Fasting brought about a progressive fall of enzyme activity. On the other hand, a rapid increase in phosphotransferase activity was elicited by giving a glucose solution to animals that had previously been fasted or fed a carbohydrate-free diet. Any route for administering glucose was effective, but neither galactose nor 2-deoxyglucose was able to do it,[2] results that agreed with the earlier studies in liver slices mentioned above.

It appeared, therefore, that the study of hexokinase would afford an opportunity to analyze at a molecular level the reasons for low hepatic glucose utilization after fasting or after feeding a glucose-free diet.

THE ISOENZYME SYSTEM FOR PHOSPHORYLATING GLUCOSE

An intriguing point in all these experiments on the effect of the amount and nature of the diet on hexokinase activity was that a residual phosphotransferase activity was always observed, even after several days of carbohydrate deprivation. At the time, this appeared puzzling in comparison with what was being observed in bacteria, where essentially no activity was detected in the absence of induction. This intriguing observation could have been explained by the existence of isoenzymes of the phosphotransferase activity, not all of which were affected by dietary manipulations. Walker's reports describing the existence in liver of two hexokinases, of which only one changed with diet,[8,9] were accordingly very well received, and were soon confirmed by the groups of Sols[10] and Weinhouse.[11] Furthermore, the activity of the diet-dependent enzyme changed also during development.[9]

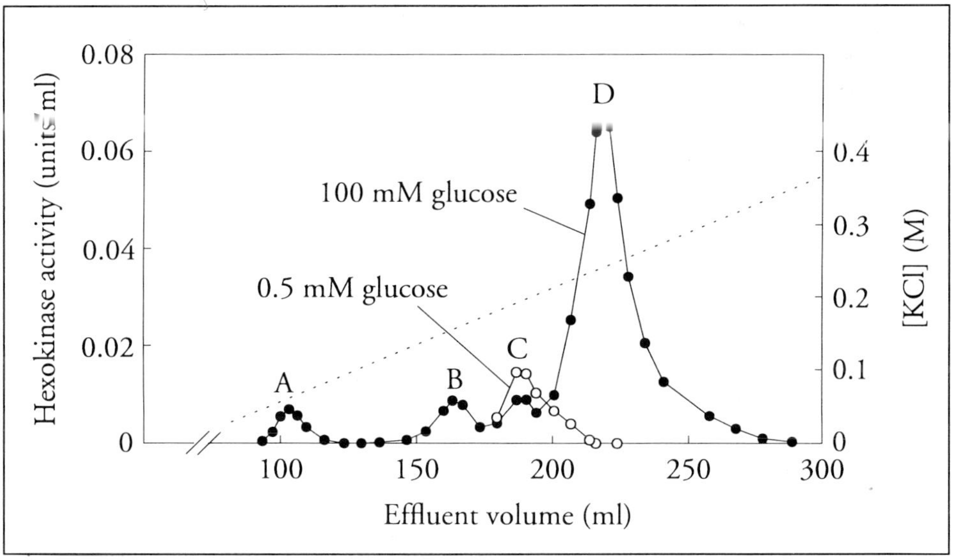

Fig. 1.1. Hexokinase isoenzymes in rat liver. Chromatography was done on DEAE cellulose in the presence of 100mM (●) or 0.5mM glucose (○), and increasing concentrations of KCl, as indicated by the broken line. Note that hexokinase C is inhibited by excess glucose at the higher concentration but not at the lower. Reproduced with permission from Biochemistry 1967; 6:460-468.

Extending the observations of these groups, Niemeyer's group demonstrated that the phosphorylation of glucose in rat liver is catalyzed not by two, but by four isoenzymes (Fig. 1.1), which they named hexokinases A, B, C and D according to their order of elution from chromatographic columns of DEAE-cellulose.[12,13] Their finding was soon confirmed on starch gel electrophoresis by Katzen et al.,[14] who named them hexokinases I, II, III, and IV (fortunately respectively). Isoenzyme D (or IV) proved to account for about 85% of the activity in the liver and to be the only one induced by carbohydrates; the other three hexokinases remained apparently unchanged with dietary modification, as has now been well established.[15] However, as we shall see in Chapter 4, most of the residual activity is due to a residual hexokinase D activity and not to the presence of the other three isoenzymes.[15] The isoenzyme D was called "glucokinase," for reasons that are presented and discussed in Chapter 2. The substrate specificity of all four isoenzymes is also analyzed there, as it is closely related to the question of whether hexokinase D is a true glucokinase or a normal hexokinase.

An important point to make at this stage is that one has to be cautious before concluding that several isoenzymes exist in any system. The simple observation that several different protein fractions catalyze the same reaction, even with some differences in kinetic parameters, is not enough to establish that they are real isoenzymes, as the apparent differences can be artifacts of the method of preparation, especially differences caused by partial proteolysis. This was of course a major problem in the early studies, but the four hexokinase isoenzymes of rat liver described by González et al. proved to be real, even though the reader of the time may have been skeptical on seeing the poor resolution of hexokinases C and D in the first paper;[12] these were better resolved in later work.[13,16]

Isoenzymes A and B are the predominant isoenzymes of the mammalian brain and muscle respectively, but both are ubiquitous[4,17,18] and they both associate with mitochondria, a localization that seems to be an important regulatory property of these isoenzymes.[19] Isoenzyme C, on the other hand, is not widely distributed, and very little is present in any tissue; however, it is abundant in ascites tumor (Novikoff hepatoma), a fact that allowed its initial characterization in mammals.[20] Its discovery in rat liver was only made possible by the use of dietary conditions that decrease the level of hexokinase D and allow the identification of hexokinase C, which is inhibited by excess glucose.[12,13]

Isoenzymes A, B and C have high affinity for glucose, that is a low K_m for this substrate, less than 0.2 mM (see Table 2.1). Hexokinase C has the highest glucose affinity of any hexokinase isoenzyme so far studied. The entire group of isoenzymes are classified as hexokinases (ATP:D-hexose 6-phosphotransferase, EC 2.7.1.1), and they have now all been purified, characterized,[20-23] cloned and sequenced.[24-30] Because of the big differences in affinity, hexokinases A, B and C are also called the high-affinity hexokinases (or low-K_m hexokinases), and hexokinase D (glucokinase) the low-affinity hexokinase. As we shall see in Chapter 3, the cooperativity of hexokinase D with respect to glucose means that it is not correct in this case to talk about a K_m or to call it the high-K_m hexokinase. Calling it a low-affinity hexokinase, on the other hand, is unexceptionable (provided that it is clear that it is glucose affinity that is referred to) as this carries no implication of Michaelis-Menten kinetics.

DIET AND ADAPTATION OF "GLUCOKINASE"

Like Niemeyer and Ureta,[1] I have used the old term "metabolic adaptation" for referring to hexokinase D activity changes in liver in response to diet, and for the same reason, namely that the change occurs in the direction that appears best to satisfy the new metabolic conditions.

Hexokinase D appeared from the very beginning to be an extremely interesting enzyme from the point of view of adaptation, not only because it was the only glucose phosphorylating isoenzyme in rat liver that varied in response to dietary modifications, the other hexokinases apparently remaining unchanged,[10-12,31] but also because its relatively low affinity for glucose compared with the other hexokinases means that it can instantaneously control the phosphorylation of glucose when

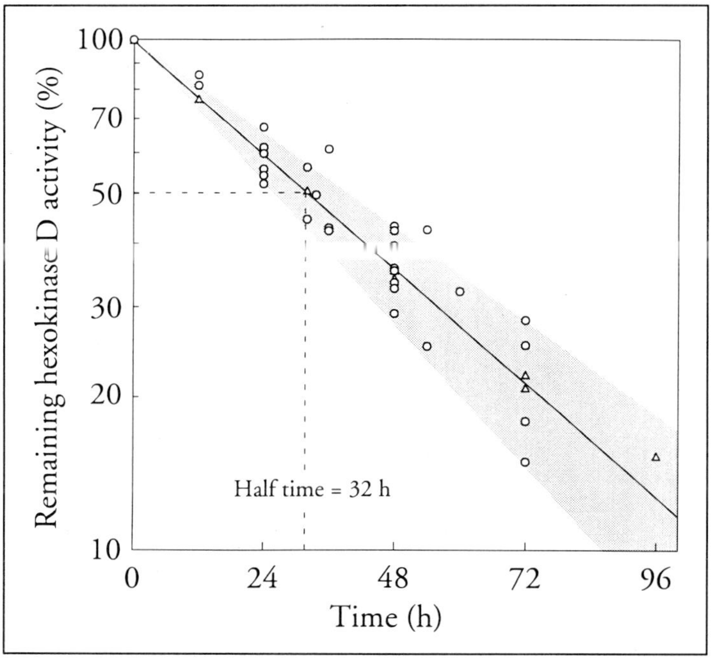

Fig. 1.2. Decay of hexokinase D activity in rat liver. Groups of rats fed a balanced diet were transferred to a carbohydrate-free diet (Δ) or were starved (O) for the indicated times, and enzyme activities per 100 g of body weight are expressed as percentages of the initial values (8.65 ± 0.57 units). Reproduced with permission from Mol Cell Biochem 1975; 6:109-126.

a load of this sugar reaches the liver, as discussed in Chapters 3 and 6, and illustrated in Figure 6.1.

Hexokinase D is thus responsible for the decrease in total glucose phosphorylating activity in response to fasting and carbohydrate-free diet; its activity also decreases in biotin-deficient animals,[32,33] even though it is not a biotin-dependent enzyme.[34] It decreased exponentially, as in a first-order process, when rats were fasted or fed a carbohydrate-free diet for several days, with a half-life of decreasing activity of about 32 hours for both conditions (Fig. 1.2).[1,2] A similar half-life has been recently obtained by Iynedjian's group,[35] confirming these early results. The kinetics of induction on the other hand appeared complex, with a transient plateau. Following a load of concentrated carbohydrate solution repeatedly given by stomach tube, there is a rapid increase of the enzyme; but after 4 to 6 hours the increase ceases during a plateau period lasting 4 to 6 hours. Afterwards the exponential increase resumes until the final level is reached at the end of 20 to 24 hours.

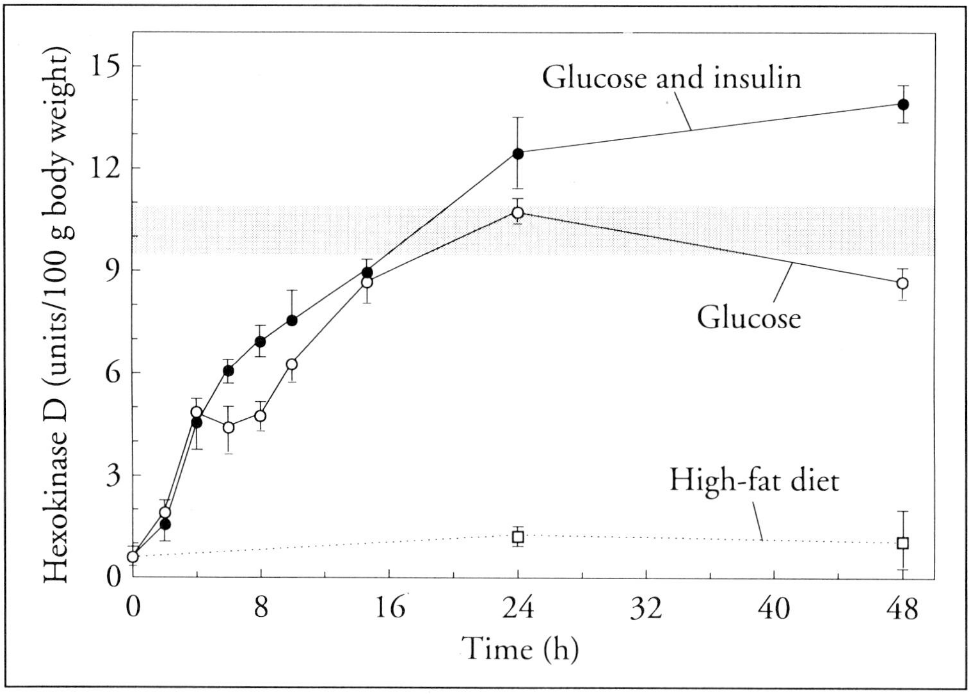

Fig. 1.3. Kinetics of glucose-mediated hexokinase D induction and the effect of exogenous insulin. Rats fed a high-fat, carbohydrate-free diet for 6 days were given a carbohydrate solution (glucose and maltodextrin) by stomach tube every 4 h until the 16th hour. Afterwards they were offered a balanced diet ad libitum. Symbols: ●, rats injected with 1 unit of crystalline insulin every 2 h until the 16th hour, when 2 units of protamine-zinc-insulin were injected; ○, rats injected with 0.15M NaCl instead of insulin; □, rats that did not receive carbohydrate and continued on the high-fat diet. Reproduced with permission from Mol Cell Biochem 1975; 6:109-126.

The duration of the plateau was decreased by the concomitant administration of exogenous insulin together with the carbohydrate (Fig. 1.3).

The induction of hexokinase D was prevented by antibiotics known to inhibit protein synthesis,[11,36,37] as well as by amino acid analogs, suggesting that the induction involved an increase in the rate of synthesis of hexokinase D, rather than an activation of already synthesized protein by covalent modification. This interpretation has now been well confirmed.[38]

HORMONAL MODULATION OF "GLUCOKINASE" INDUCTION

The contribution of the endocrine system to enzyme adaptation appears to be a consistent feature of multicellular organisms. Changes in the diet, or in other parameters of the environment, evoke alterations in the balance of the endocrine system, promoting secretion of some hormones and blocking that of others. The endocrine system accordingly plays an important role in hexokinase D adaptation, as with other mammalian enzymes, and from the early studies it was shown that several hormones contribute to modulate the hexokinase D levels.

Insulin

An important discovery made quite early was that insulin is essential both for maintaining hexokinase D levels and for inducing the enzyme after its depletion. For example, DiPietro and Weinhouse demonstrated that glucose phosphorylating activity decreased in the liver of alloxan-diabetic rats.[39] Two years later, it was recognized that hexokinase D or the low-affinity isoenzyme was responsible for the changes, because it decreased in the diabetic rat and recovered normal values after the injection of insulin to animals fed *ad libitum*.[10,11,36] These observations were soon confirmed and expanded in other laboratories.[31,40-42]

The requirement of insulin for the induction of hexokinase D was clearly demonstrated in experiments by Niemeyer's group, who showed that the administration of anti-insulin immune serum inhibited the induction of hexokinase D in intact animals fed a carbohydrate-free diet (Fig. 1.4).[42] In agreement with this result, it was later shown that the increase of hexokinase D evoked by feeding glucose after starvation could be prevented by injecting mannoheptulose, a sugar that suppresses the release of insulin from the pancreas.[43] The effects of both the antiserum and mannoheptulose were counteracted by exogenous insulin.

Pyruvate kinase and phosphofructokinase were also decreased in the diabetic rat, and recovered to normal values after insulin administration. This suggested at the time the possibility of a common operon, but the responses of the three enzymes are uncoupled when alloxan-diabetic rats are fed a fructose-glycerol diet,[44] as has been confirmed in more recent experiments (see Chapter 8).

Although hexokinase D is the only isoenzyme in liver whose activity depends on insulin,[15] this is not the case in other tissues, as we shall see in Chapter 4, and insulin also contributes to the regulation of glucose phosphorylation in tissues not containing hexokinase D. Diabetes results in a decreased level of isoenzyme B in insulin-sensitive tissues, such as muscle,[45,46] adipose tissue[43,46,47] and mammary gland.[48]

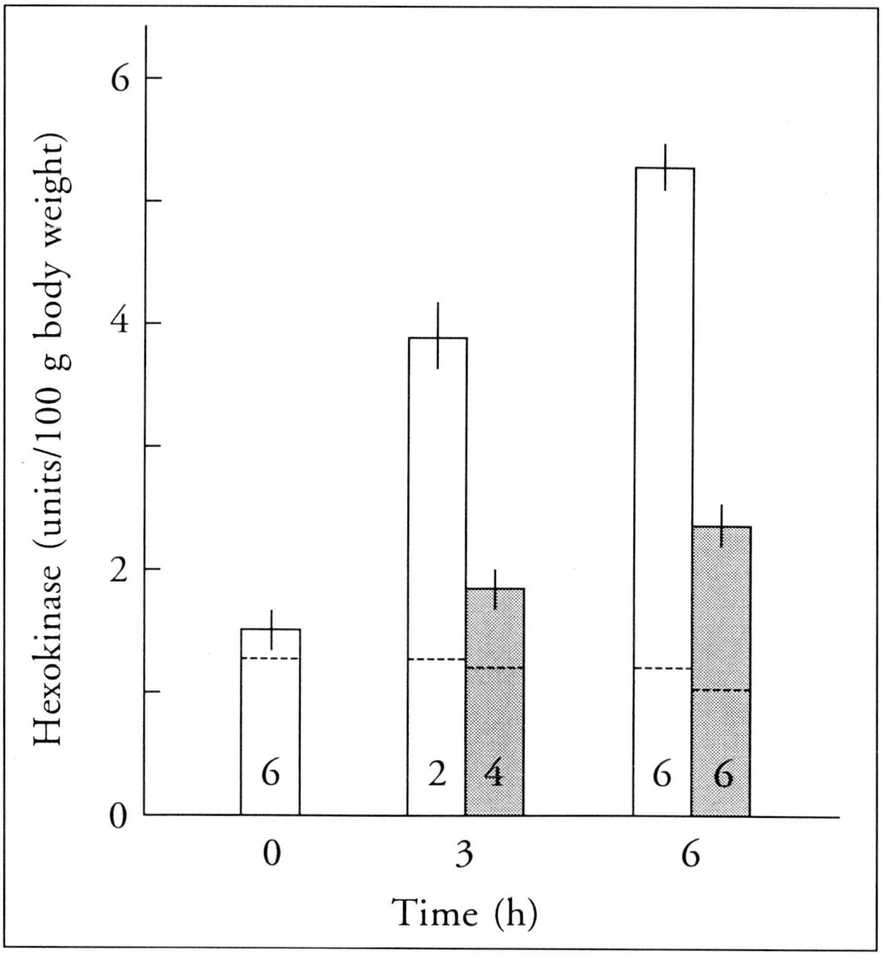

Fig. 1.4. Effect of insulin antiserum on the substrate induction of hexokinase D in rat liver. Rats previously fed a high-fat, carbohydrate-free diet for 6 days were subjected to various treatments before being killed for enzyme assay. Ten rats (shaded bars) received intravenously 0.5-0.7 ml of anti-insulin guinea pig serum and 15 min later a solution of carbohydrate by stomach tube. Four of these animals were killed 3 h after administration of carbohydrate. The other six received a second dose of immune serum and carbohydrate at this time and were killed 3 h later. Groups of rats similarly treated in which normal guinea pig serum replaced the immune serum acted as controls (unshaded bars). A group of animals did not receive any treatment and acted as initial control. Bars represent total glucose phosphorylating activity measured with 100 mM glucose, and the dotted lines indicate parallel assays with 0.5 mM glucose (at which concentration the activities of hexokinases A, B and C are high and that of hexokinase D negligible). The figures in the bars show the numbers of rats in each group. Reproduced with permission from J Biol Chem 1967; 242:860-864.

Glucocorticoids

Contrary to the case of insulin, the adrenal glands are essential neither for the maintenance of normal levels of liver hexokinase D nor for the induction of the enzyme after depletion by fasting or by a carbohydrate-free diet.[31,37] However, glucocorticoids seem to modulate the rate of hexokinase D increase,[37] as the rate of induction evoked by the administration of glucose to animals fed a carbohydrate-free diet is lower in adrenalectomized than in sham-operated rats, and it can be brought to normal values by the administration of cortisol. In addition, injection of glucocorticoids permits the precocious appearance of hexokinase D in newborn rats.[49]

Pituitary gland

The pituitary gland also appears to exert a modulatory action on hexokinase D adaptation, although the results were initially unclear: some authors described a decrease in liver hexokinase D after hypophysectomy, whereas others observed no effect. However, normal values of hexokinase D were maintained when approximate physiological doses of cortisol, thyroxine and testosterone were given to the surgically treated animals to compensate for the atrophy of the corresponding secreting glands. Although the final levels of hexokinase D were normal in hypophysectomized rats supplied with cortisol, thyroxine and testosterone, the rate of induction was slower than in the sham-operated controls, which suggests a modulating role for the pituitary gland.

THE INDUCER: GLUCOSE OR INSULIN?

A problem apparent in the earliest studies was whether the true inducer was glucose or insulin, or if both were equally required to promote the hexokinase D increase. The problem was difficult, as glucose was known to stimulate insulin release[50-52] and low levels of plasma insulin were observed in rats deprived of food[53] and in rats fed a high-fat diet.[54] In an attempt to solve this problem it was shown that insulin alone (in different doses and schedules of administration) was unable to increase hexokinase D previously decreased by a carbohydrate-free diet;[41] nor could it prevent or modify the decay of hexokinase D resulting from feeding the carbohydrate-free diet.[2] All of these observations led to the conclusion that induction of hexokinase D required both glucose and insulin. (Nowadays this result would be easily explained by the capacity of fasting and a carbohydrate-free diet to stimulate release of glucagon, which inhibits induction of hexokinase D, as discussed below.)

Actinomycin D caused a complete halt to hexokinase D increase when administered to animals at the beginning of the induction, but only a slight decrease in the rate of induction if it was injected after several hours.[37,55] The same type of response to actinomycin was observed in alloxan-diabetic rats fed a carbohydrate-free diet when it was supplied with both glucose and insulin.[2] This difference in the response to actinomycin according to the time of injection was interpreted as the consequence of a long-lived messenger RNA formed during the first hours of induction in the absence of the antibiotic.[37] In experiments with diabetic rats in which either glucose or insulin was supplied at zero time, the injection 4 hours later of actinomycin with carbohydrate and insulin blocked the subsequent induction of hexokinase D completely. These results seemed to indicate that in contrast to the case with both glucose and insulin present, neither of them alone produced an accumulation of messenger RNA active in the second 4-hour period, when actinomycin was present. These observations stress the necessity of both glucose and insulin for the induction of hexokinase D and suggest that the nutrient and the hormone act together at the transcriptional level. However, unequivocal evidence was lacking in these early studies, and thus some investigators argued that insulin was the inducer,[36,55,56] whereas others ascribed a rather permissive role to insulin.[41] This point was only solved about two decades afterwards, when it was clearly established that insulin is the real inducer,[38] as we shall see in Chapter 8.

INHIBITION OF "GLUCOKINASE" INDUCTION BY GLUCAGON AND CATECHOLAMINES

Glucagon inhibits the induction of hexokinase D, and may even block it completely, depending on the doses used.[41,43,57] As in the case of actinomycin, the effect depends on the time of administration. Thus a dose that completely blocks the induction when given at the start of the experiment inhibits the enzyme increase only slightly if it is given after several hours of induction (Fig. 1.5a).

Catecholamines (epinephrine, norepinephrine, isoproterenol) inhibit hexokinase D induction,[58-60] and their effect is very similar to that of glucagon with respect to the time of administration. Thus, the inhibition is complete if the hormones are administered at the initiation of the induction, but only partial if they are injected after the induction is already in progress (Fig. 1.5b).[60]

The modality of the response suggested that the hormones act at the initial steps of the induction and that once the synthesis of hexokinase D has started the process becomes rather insensitive to the hormonal action. As the effects of both types of hormones, glucagon and catecholamines, mimic that of actinomycin D, it was postulated that they also act at the transcriptional level, even though different mechanisms may operate.[1,2,58,60]

As expected from these results, cyclic AMP, but not AMP, inhibits hexokinase D induction after the administration of glucose to rats fed the carbohydrate-free diet, the effect being dependent on the dose injected.[60]

absent from birds, all reptiles so far studied, some leptodactylid frogs and all bufonid toads. Hexokinase A has been unambiguously identified in turtles and in the frogs that lack hexokinase C. The lack of unique properties to identify hexokinase B unambiguously has prevented its taxonomic distribution from being similarly established (it is defined in practice by the fact that it is not hexokinase A, C or D).

It appears that the hexokinase isoenzyme system has conserved its main features throughout the vertebrates apart from birds and higher reptiles, whose hexokinases differ from those of mammals, turtles and amphibians: this is discussed in two articles from Ureta's group.[77,78]

In contrast to the kinetic cooperativity of vertebrate hexokinase D, its variation in response to diet is not a conserved feature. Thus, although hexokinase D activity decreases after fasting in rat, mouse and hamster, there is no significant change in the guinea pig and the degu.[73] In addition, different strains of mouse respond differently to diet; a carbohydrate-free diet (high protein, high fat) causes a decrease of hexokinase D activity in mice of strain C57BL but not in those of the BALB type. Furthermore, in contrast to what occurs in the rat, carbohydrate alone is unable to sustain normal activities of hexokinase D in both strains of mouse.[73]

Among the other species with adaptive hexokinase D are dog[79] and human.[80] By contrast, amphibians such as the toad show no change of the hexokinase D activity after several days of fasting.[1] In turtle, there are suggestions that the liver enzyme disappears during hibernation,[81] but no systematic studies have been done.

PERSPECTIVES

In the light of what is now known about the molecular mechanism of induction (see Chapter 8) it will be very interesting to clone the gene of a species that does not respond to fasting, and to examine the promoter. For this purpose the two strains of mice mentioned above, C57BL and BALB, offer an especially valuable tool, and the case of the mouse deserves further comments for other reasons. First of all, the regulation of hexokinase D activity is less well understood in the mouse than in the rat. There are genetic differences that affect basal levels of the enzyme in a wide variety of inbred strains. Coleman suggested that alleles of a hexokinase D gene confer high activity to GK^a/GK^a mice and low activity to GK^b/GK^b mice.[82] Walker's group confirmed the existence of alleles of this gene in strains C3H/He (high hexokinase D activity, GK^a/GK^a genotype) and C58 (low hexokinase D activity, GK^b/GK^b genotype), and showed that alleles of the gene segregate in a single codominant manner in crosses between these two strains.[83] Studies of partially purified hexokinase D from the two strains failed to reveal any structural difference between the two enzymes, as judged by kinetic, electrophoretic and immunological properties.[84] The immunological studies showed, however, that the amounts of enzyme were different in the two strains, so that the high-activity C3H/He strain had twice as many immunoreactive hexokinase D molecules as the low-activity C58 strain. The same group showed later that the mRNA levels are larger in strains with high enzyme activity than in those with low enzyme activity, suggesting that the strains differ either in transcription of the hexokinase D gene or in the stability of the mRNA.[85] It will be especially interesting to examine the controlling sequences around the hexokinase D gene in these two strains, and to compare them with those of the BALB strain, which appears to have an intermediate level of enzyme activity.

References

1. Niemeyer H, Ureta T. Enzyme adaptation in mammals. In: Molecular Basis of Biological activity (PAABS Symposium. vol. 1). New York: Academic Press, 1972:221-272.
2. Niemeyer H, Ureta T, Clark-Turri L. Adaptive character of liver glucokinase. Mol Cell Biochem 1975; 6:109-126.
3. Walker DG. The nature and function of hexokinases in animal tissues. Essays Biochem 1966; 2:33-37.
4. Weinhouse S. Regulation of glucokinase in liver. Curr Topics Cell Regul 1976; 11:1-50.
5. Niemeyer H, Clark-Turri L, Garcés E et al. Selective response of liver enzymes to the administration of different diets after fasting. Arch Biochem Biophys 1962; 98:77-85.
6. Niemeyer H, Clark-Turri L, Rabajille E. Induction of glucokinase by glucose in rat liver. Nature 1963; 198:1096-1097.
7. Pérez N, Clark-Turri L, Rabajille E, Niemeyer H. Regulation of rat liver enzymes by natural components of the diet. J Biol Chem 1964; 239:2420-2426.
8. Walker DG. The development of hepatic hexokinases after birth. Biochem J 1962; 84:118P-119P.
9. Walker DG. On the presence of two soluble glucose-phosphorylating enzymes in adult liver and the development of one of these after birth. Biochim Biophys Acta 1963; 77:209-266.
10. Viñuela E, Salas M, Sols A. Glucokinase and hexokinase in liver in relation to glycogen synthesis. J Biol Chem 1963; 238:PC1175-1177.
11. Sharma C, Manjeshwar R, Weinhouse S. Effects of diet and insulin on glucose-adenosine triphosphate phosphotransferases of rat liver. J Biol Chem 1963; 238:3840-3845.
12. González C, Ureta T, Sánchez R et al. Multiple molecular forms of ATP:hexose 6-phosphotransferase from rat liver. Biochem Biophys Res Commun 1964; 16:347-352.
13. González C, Ureta T, Babul J et al. Characterization of isoenzymes of adenosine triphosphate:D-hexose 6-phosphotransferase from rat liver. Biochemistry 1967; 6:460-468.

14. Katzen HM, Soderman DD, Nitowski HM. Kinetic and electrophoretic evidence for multiple forms of glucose-ATP phosphotransferase activity from human cell cultures and rat liver. Biochem Biophys Res Commun 1965; 19:377-382.
15. Reyes A, Rabajille E, Cárdenas ML et al. Stability of hexokinases A, B, and C, and *N*-acetylglucosamine kinase in liver cells isolated from rats submitted to diabetes and several dietary conditions. Biochem J 1984; 221:311-315.
16. Grossbard L, Schimke RT. Multiple hexokinases of rat tissues. Purification and comparison of soluble forms. J Biol Chem 1966; 241:3546-3560.
17. Colowick SP. The hexokinases. In: Boyer PD, ed. The Enzymes. 3rd ed. New York: Academic Press, 1973; 9:1-48.
18. Ureta T. The comparative isozymology of vertebrate hexokinases. Comp Biochem Physiol 1982; 71B:549-555.
19. Wilson JE. Regulation of mammalian hexokinase activity. In: Beiner R, ed. Regulation of carbohydrate metabolism. Boca Ratón: CRC Press, 1985; 1:45-85.
20. Radojkovic J, Ureta T. Hexokinase isoenzymes from the Novikoff hepatoma. Purification, kinetic and structural characterization, with emphasis on hexokinase C. Biochem J 1987; 242:895-903.
21. Chou AC, Wilson JE. Purification and properties of rat brain hexokinase. Arch Biochem Biophys 1972; 151:48-55.
22. Holroyde MJ, Trayer IP. Purification and properties of rat skeletal muscle hexokinase. FEBS Lett 1976; 62:215-219.
23. Holroyde MJ, Allen MB, Storer AC et al. The purification in high yield and characterization of rat hepatic glucokinase. Biochem J 1975; 153: 363-373.
24. Nishi S, Seino S, Bell GI. Human hexokinase: sequences of amino- and carboxyl-terminal halves are homologous. Biochem Biophys Res Comm 1988; 157:937-943.
25. Schwab DA, Wilson JE. Complete amino acid sequence of rat brain hexokinase, deduced from the cloned cDNA, and proposed structure of a mammalian hexokinase. Proc Natl Acad Sci USA 1989; 86:2563-2567.
26. Thelen AP, Wilson JE. Complete amino acid sequence of the type II isozyme of rat hexokinase, deduced from the cloned cDNA: comparison with a hexokinase from Novikoff ascites tumor. Arch Biochem Biophys 1991; 286:645-651.
27. Schwab DA, Wilson JE. Complete amino acid sequence of the type III isoenzyme of rat hexokinase, deduced from the cloned cDNA. Arch Biochem Biophys 1991; 285:365-370.
28. Andreone TL, Printz RL, Pilkis SJ et al. The amino acid sequence of rat liver glucokinase deduced from cloned cDNA. J Biol Chem 1989; 264:363-369.
29. Magnuson MA, Shelton KD. An alternate promoter in the glucokinase gene is active in the pancreatic β-cell. J Biol Chem 1989; 264:15936-15942.
30. Nishi S, Stoffel M, Xiang K et al. Human pancreatic beta-cell glucokinase: cDNA sequence and localization of the polymorphic gene to chromosome 7, band p 13. Diabetologia 1992; 35:743-747.

31. Walker DG, Rao S. The role of glucokinase in the phosphorylation of glucose by rat liver. Biochem J 1964; 90:360-368.
32. Dakshinamurti K, Cheah-Tan C. Liver glucokinase of the biotin deficient rat. Can J Biochem 1968; 46:75-80.
33. Dakshinamurti K, Cheah-Tan C. Biotin-mediated synthesis of hepatic glucokinase in the rat. Arch Biochem Biophys 1968; 127:17-21.
34. Salas J, Viñuela E, Salas M et al. Glucokinase of rabbit liver. Purification and properties. J Biol Chem 1965; 240:1014-1018.
35. Iynedjian PB, Gjinovci A, Renold AE. Stimulation by insulin of glucokinase gene transcription in liver of diabetic rats. J Biol Chem 1988; 263:740-744.
36. Salas M, Viñuela E, Sols A. Insulin-dependent synthesis of liver glucokinase in the rat. J Biol Chem 1963; 238:3535-3538.
37. Niemeyer H, Clark-Turri L, Pérez N et al. Studies of factors affecting the induction of ATP:D-hexose 6-phosphotransferase in rat liver. Arch Biochem Biophys 1965; 109:634-645.
38. Iynedjian PB. Mammalian glucokinase and its gene. Biochem J 1993; 293:1-13.
39. DiPietro DL, Weinhouse S. Hepatic glucokinase in the fed, fasted and alloxan-diabetic rat. J Biol Chem 1960; 235:2542-2445.
40. Blumenthal MD, Abraham S, Chaikoff IL. Dietary control of liver glucokinase activity in the normal rat. Arch Biochem Biophys 1964; 104:215-224.
41. Niemeyer H, Pérez N, Rabajille E. Interrelation of actions of glucose, insulin, and glucagon on induction of adenosine triphosphate: D-hexose phosphotransferase in rat liver. J Biol Chem 1966; 241:4055-4059.
42. Niemeyer H, Pérez N, Codoceo R. Liver glucokinase induction in acute and chronic insulin insufficiency in rats. J Biol Chem 1967; 242:860-864.
43. Pilkis SJ. Hormonal control of hexokinase activity in animal tissues. Biochim Biophys Acta 1970; 215:461-476.
44. Sillero A, Sillero MAG, Sols A. Regulation of the level of key enzymes of glycolysis and gluconeogenesis in liver. Eur J Biochem 1969; 10:351-354.
45. Grossbard L, Weksler M, Schimke RT. Electrophoretic properties and tissue distribution of multiple forms of hexokinase in various mammalian species. Biochem Biophys Res Commun 1966; 24:32-38.
46. Katzen HM, Soderman DD, Wiley CE. Multiple forms of hexokinase: activities associated with subcellular particulate and soluble fractions of normal and streptozotocin diabetic rat tissues. J Biol Chem 1970; 245:4081-4096.
47. Katzen HM. The multiple forms of mammalian hexokinase and their significance to the action of insulin. Adv Enz Reg 1967; 5:335-356.
48. Walters E, McLean P. Multiple forms of glucose-adenosine triphosphate phosphotransferase in rat mammary gland. Biochem J 1967; 104:778-783.
49. Jamdar SC, Greengard O. Premature formation of glucokinase in developing rat liver. J Biol Chem 1970; 245:2779-2783.
50. Grodsky GM, Batts AA, Bennett LL et al. Effects of carbohydrates on secre-

tion of insulin from isolated rat pancreas. Am J Phys 1963; 205:638-644.
51. Coore HG, Randle PJ. Regulation of insulin secretion studied with pieces of rabbit pancreas incubated in vitro. Biochem J 1964; 93:66-78.
52. Karam JH, Grasso SG, Wegienka LC et al. Effect of selected hexoses, of epinephrine and of glucagon on insulin secretion in man. Diabetes 1966; 15:571-578.
53. Hales CN, Randle PJ. Immunoassay of insulin with insulin-antibody precipitate. Biochem J 1963; 93:137-146.
54. Blázquez E, Lopez Quijada C. The effect of a high-fat diet on glucose, insulin sensitivity and plasma insulin in rats. J Endocrinol 1968; 42:489-494.
55. Sols A, Sillero A, Salas J. Insulin-dependent synthesis of glucokinase. J Cell Physiol 1965; 66 (suppl. 1):23-38.
56. Weber G, Singhal RL, Stamm NB et al. Synchronous behavior pattern of key glycolytic enzymes: glucokinase, phosphofructokinase, and pyruvate kinase. Adv Enz Reg 1966; 4:59-81.
57. Pitot HC, Peraino C, Pries N et al. Template stability in liver and hepatoma. Adv Enz Reg 1965; 3:359-368.
58. Niemeyer H, Pérez N, Rabajille E. Interrelation of actions of glucose, insulin, and glucagon on induction of adenosine triphosphate:D-hexose phosphotransferase in the rat. J Biol Chem 1966; 241:4055-4059.
59. Pitot HC, Peraino C, Pries N et al. Glucose repression and induction of enzyme synthesis in the liver. Adv Enz Reg 1964; 2:237-247.
60. Ureta, T, Radojkovic J, Niemeyer H. Inhibition by catecholamines of the induction of rat liver glucokinase. J Biol Chem 1970; 245:4819-4824.
61. Clark-Turri L, Peñaranda J, Rabajille E et al. Immunochemical titration of liver glucokinase from normal, fasted, and diabetic rats. FEBS Lett 1974; 41:342-344.
62. Sibrowski W, Staegemann U, Seitz HJ. Accelerated turnover of hepatic glucokinase in starved and streptozotocin-diabetic rat. Eur J Biochem 1982; 127:571-574.
63. Katz NR, Nauck MA, Wilson PT. Induction of glucokinase by insulin under the permissive action of dexamethasone in primary rat hepatic cultures. Biochem Biophys Res Comm 1979; 88:23-29.
64. Schudt C. Hormonal regulation of glucokinase in primary cultures of rat hepatocytes. Eur J Biochem 1979; 98:77-82.
65. Nakamura T, Aoyama K, Ichihara A. Precocious induction of glucokinase in primary cultures of postnatal rat hepatocytes. Biochem Biophys Res Comm 1979; 91:513-520.
66. Spence JT, Pitot HC. Hormonal regulation of glucokinase in primary cultures of adult rat hepatocytes. J Biol Chem 1979; 254:12331-12336.
67. Wakelam MJO, Walker DG. De novo synthesis of glucokinase in hepatocytes from neonatal rats. FEBS Lett 1980; 111:115-119.
68. Spence JT, Merrill MJ, Pitot HC. Role of insulin, glucose, and cyclic GMP in the regulation of glucokinase in cultured hepatocytes. J Biol Chem 1981; 256:1598-1603.
69. Iynedjian PB, Pilot P-R, Nouspikel T et al. Differential expression and

regulation of the glucokinase gene in liver and islets of Langerhans. Proc Natl Acad Sci USA 1989; 86:7838-7842.

70. Spence JT. Levels of translatable mRNA coding for rat liver glucokinase. J Biol Chem 1983; 258:9143-9146.
71. Sibrowski W, Seitz HJ. Rapid action of insulin and cyclic AMP in the regulation of functional messenger RNA coding for glucokinase in rat liver. J Biol Chem 1984; 259:343-346.
72. Minderop RH, Hoeppner W, Seitz HJ. Regulation of hepatic glucokinase gene expression. Eur J Biochem 1987; 164:181-187.
73. Ureta T, González C, Lillo S, Niemeyer H Comparative studies on glucose phosphorylating isoenzymes of vertebrates I. The influence of fasting and the nature of the diet on liver glucokinase and hexokinases of rodents. Comp Biochem Physiol 1971; 40B:71-80.
74. Ureta T, Radojkovic J, Zepeda S et al. Comparative studies on glucose phosphorylating isoenzymes of vertebrates. VII. Mammalian hexokinases. Comp Biochem Physiol 1981; 70B:225-236.
75. Ureta T. The allosteric regulation of hexokinase C from amphibian liver. J Biol Chem 1976; 251:5035-5042.
76. Ureta T, Radojkovic J, Díaz N et al. Comparative studies on glucose phosphorylating isoenzymes of vertebrates. Identification and characterization of amphibian liver hexokinases. Arch Biochem Biophys 1978; 186:235-247.
77. Ureta T, Radojkovic J, Preller A et al. Glucose utilization in vertebrates as a molecular probe for the study of evolution. Arch Biol Med Exp 1979; 12:49-58.
78. Ureta T, Radojkovic J, Lagos R et al. Phylogenetic and ontogenetic studies of glucose phosphorylating isoenzymes of vertebrates. Arch Biol Med Exp 1979; 12:587-609.
79. Hornichter R, Brown J, Snow H. Effects of starvation and diabetes on glucokinase activity in dog liver. Clin Res 1967; 15:109.
80. Borrebaek B, Hultman E, Nilsson LH et al. Adaptable glucokinase activity of human liver. Biochem Med 1970; 4: 469-475.
81. Cárdenas ML, Rabajille E, Niemeyer H. Kinetic cooperativity of glucokinase with glucose. Arch Biol Med Exp 1979; 12:571-580.
82. Coleman DL. Genetic control of hexokinase activity in mice. Biochem Genet 1977; 15:297-305.
83. Lavender FL, James PA, Walker DG. Hepatic glucokinase activity and circulating insulin concentrations in two inbred mouse strains. Diabetologia 1983; 25:114-119.
84. James PA, Lavender FL, Lawrence GM et al. Comparison of the glucokinase in C3H/He and C58 mice that differ in their hepatic activity. Biochem Genet 1985; 23:525-538.
85. Middleton RJ, Walker DG. A comparison of hepatic glucokinase gene expression in high-and low-activity strains of mice. Comp Biochem Physiol 1992; 102B:337-342.

CHAPTER 2

Glucokinase or Hexokinase?

This chapter describes the general characteristics of the reaction catalyzed by the hexokinases, as well as the historical background that led to the use of the name of "glucokinase" for hexokinase D or IV. As discussed below, this name is misleading, and that is why I have put it in quotation marks in the title of the book. There are several articles[1-3] in which more detailed arguments may be found.

THE REACTION CATALYZED BY HEXOKINASES

Before addressing the question of whether the isoenzyme characteristic of the liver is better called glucokinase or hexokinase, I need to describe the hexokinase reaction, so as to have a better basis for making a judgement.

The term "hexokinase" was introduced by Meyerhof[4] in 1927 to refer to a preparation from a yeast autolysate that catalyzed the phosphorylation of glucose by ATP. Similar enzymic activity was subsequently found in muscle tissue and later in other types of cells. Crane[5], who gives references to the earlier work, classified hexokinases into single-substrate and multisubstrate enzymes. All the mammalian enzymes, and hexokinase D is no exception, come into the second category, as they all are capable of phosphorylating a number of different acceptor molecules (see Table 2.2, below), and binding others that can act as inhibitors even if they are not substrates. In every case, glucose is the principal physiological substrate[6] and has the highest phosphorylation coefficient, as will be discussed below; again, hexokinase D is not an exception.

In terms of chemistry, hexokinases catalyze the phosphorylation of various hexoses (not just glucose) by ATP according to the following equation:

$$R - CH_2OH + MgATP^{2-} \rightarrow R - CH_2O - PO_3^{\ 2-} + MgADP^{-} + H^{+}$$

The acceptor molecule ($R\text{-}CH_2OH$) is the cyclic form of a hexose such as glucose, mannose, fructose, or 2-deoxyglucose. The reaction occurs at C6 (in all substrates except fructose this is the only CH_2OH group in the molecule), even though each substrate molecule carries several other OH groups. Magnesium ions are necessary for the reaction because the true nucleotide substrate is the $MgATP^{2-}$ ion. The hexokinases, including hexokinase D, can use analogs of ATP such as ITP, but the rate is very low[6,7] and in general the nucleotide specificity is stricter than the sugar specificity.

The equilibrium for the hexokinase reaction strongly favors formation of ADP. A proton appears in the reaction, and so the equilibrium position varies with pH. Under physiological conditions, the phosphorylation of glucose by ATP in the presence of hexokinase has a standard Gibbs energy of the order of -20 kJ/mol (-5 kcal/mol), and it is therefore essentially irreversible.

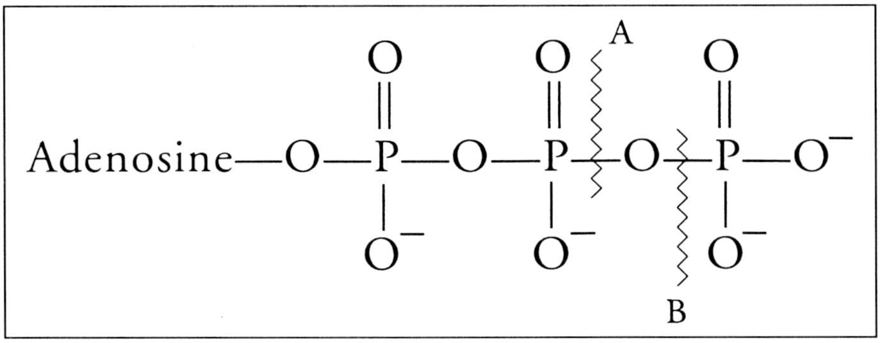

Fig. 2.1. Scission of ATP in the hexokinase reaction. Although in principle one could imagine transfer of either a phospho group (scission at A) or a phosphoryl group (scission at B), ^{18}O tracer studies have shown that only the phosphoryl group is transferred.

In terms of the complete chemical reaction the result would be the same regardless of whether the complete -OH group of the accepting alcohol was replaced by a complete phosphate (or phospho) group $-OPO_3^{2-}$, or whether only the H atom of the alcohol was replaced by a phosphoryl group $-PO_3^{2-}$, as illustrated in Figure 2.1. In fact the group transferred is a phosphoryl and not a phosphate group, as was shown by Cohn[8] for yeast hexokinase in the presence of $H_2{}^{18}O$; she found that no ^{18}O could be detected in either of the products, ADP and glucose 6-phosphate, a result consistent only with phosphoryl transfer. Nonetheless, the enzymes continue to be called phosphotransferases (rather than phosphoryltransferases), in keeping with the principle commonly applied in enzyme nomenclature that enzymes are named according to the complete reaction catalyzed, usually without taking account of mechanistic details. Although this principle sometimes has unfortunate consequences it avoids the need for constant renaming as more mechanistic information becomes available.

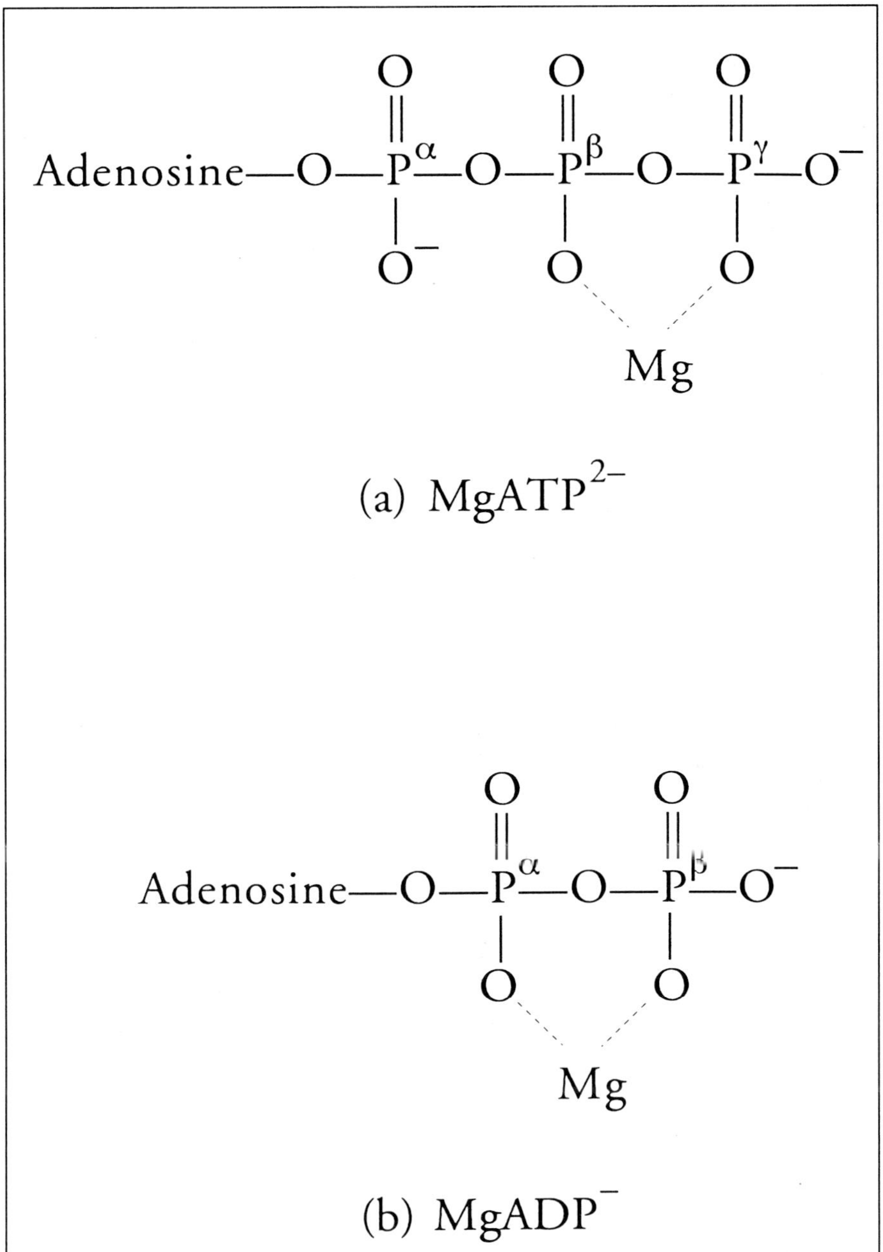

Fig. 2.2. Binding of magnesium ion to adenine nucleotides in free solution. (a) In $MgATP^{2-}$ the Mg^{2+} ion binds principally between the β and γ phosphates; (b) in $MgADP^{-}$ it binds between the α and β phosphates.

The reaction is a nucleophilic attack by the acceptor alcohol on the terminal phosphate group of ATP. Nuclear magnetic resonance studies[9] on the behavior of nucleotides and metal ions in solution suggested that Mg^{2+} complexes with ATP^{4-} principally by interacting with oxygen atoms on the β and γ phosphates, as illustrated in Figure 2.2a, and similarly with ADP^{3-} (Fig. 2.2b). By complexing with these atoms the Mg^{2+} ion reduces the electron density around the two phosphorus atoms, thereby increasing the ease with which a nucleophile can attack the terminal phosphorus atom. The phosphoryl transfer reaction may be visualized as the breaking of one O-P bond and the making of another. An obvious possible mechanism would be the formation of a covalently linked phosphoryl-enzyme intermediate, but no evidence for this has been forthcoming. On the other hand, formation of an intermediate phosphoryl-enzyme at very low concentration is unlikely from thermodynamic considerations.

UNITY OF MECHANISM IN THE HEXOKINASE REACTION

Stereochemical course

The currently accepted view is that the mechanism involves a direct transfer of the phosphoryl group from ATP to the acceptor molecule, which is supported by studies with adenosine 5'-[γ(*S*)-^{16}O,^{17}O,^{18}O]triphosphate to determine the stereochemical course of phosphoryl transfer catalyzed by yeast hexokinase. In principle, there are four possible mechanisms for transferring a phosphoryl group during the interconversion of intermediates in a kinase-catalyzed reaction:

(i) The *associative "in line" mechanism*, similar to a Walden inversion in an S_N2 reaction, in which the new bond begins to be made before the old one is broken;

(ii) The *dissociative mechanism*, in which the transferred group dissociates (as metaphosphate) before the new bond is formed, as in an S_N1 reaction;

(iii) The *adjacent mechanism*, with pseudorotation (or "turnstile" rotation); and

(iv) The *double-displacement mechanism*, with a transient phosphoryl-enzyme intermediate, which means two transfers, first to the enzyme and second from the enzyme to the acceptor.

The first mechanism occurs with inversion of configuration at phosphorus. It seems likely that the second mechanism also occurs with inversion of configuration (in contrast to the corresponding S_N1 reaction in free solution, in which there is racemization) because the enzyme must bind the metaphosphate intermediate tightly to prevent

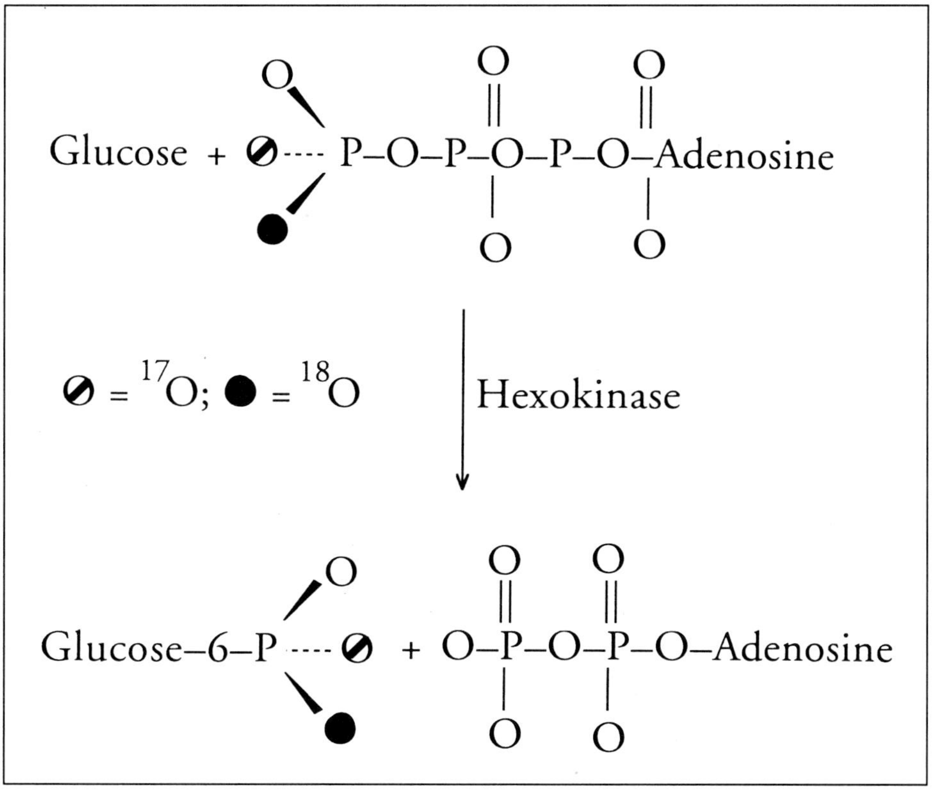

Fig. 2.3. Inversion of configuration during the hexokinase reaction. Use of ^{17}O and ^{18}O as labels allows the stereochemical course of the hexokinase reaction to be determined, and shows that there is inversion of configuration at the transferred P atom.

it being captured by solvent water; however, the possibility of partial (or total) racemization cannot be excluded. The third mechanism should lead to retention of configuration, as should the fourth mechanism, if the reasonable assumption is made that the two phosphoryl-transfer steps follow the same stereochemical course.

The analysis of the chirality of the product, D-glucose 6-[^{16}O,^{17}O,^{18}O] phosphate by ^{31}P n.m.r. spectroscopy has shown that in yeast hexokinase the reaction proceeds with inversion of configuration at phosphorus,[10] which can be visualized as illustrated in Figure 2.3. This result indicates an associative "in-line" mechanism with direct transfer of the phosphoryl group between substrates in the ternary complex. The alternative dissociative mechanism with metaphosphate as an intermediate has been excluded by positional isotope-exchange measurements.[11] One cannot rule out triple-displacement mechanisms similar to the double-displacement mechanism but involving two (or a larger even number) of different phosphoryl-enzymes; most investigators have assumed these to be unnecessarily complicated and hence improbable, though Spector has forcefully argued an alternative view.[12] Furthermore, the observation that the Λ isomer of the β,γ-bidentate $Co(NH_3)_4ATP$ complex is a substrate for yeast hexokinase suggests that Mg^{2+} in the ternary complex between hexokinase, glucose and $MgATP^{2-}$ is bound to the β,γ-phosphate residues of ATP,[13] which is consistent with the associative mechanism (Fig. 2.4).

Although the most detailed studies have been done with yeast hexokinase (not always with a clearly identified isoenzyme) the stereochemical course of the reaction catalyzed by liver hexokinase D has been studied,[14] with essentially the same result as that found with yeast hexokinase, supporting the view that the two enzymes (and probably all the other hexokinases) follow the same chemical mechanism.

Stereospecificity for the metal-nucleotide binding site

Hexokinase D and the other hexokinases appear to have a similar stereospecificity for the metal-nucleotide binding site. Diastereoisomers of the nucleoside phosphorothioates (ATPβS) have been used as probes of the structure of the metal nucleotide substrate and of the nucleotide binding site of yeast hexokinase.[15] There are two diastereoisomers of ATPβS, A and B, which can be distinguished on the basis of their activity in the presence of Mg^{2+}. Each of them can exist in two forms, Λ and Δ, as shown in Figure 2.5. As Mg^{2+} coordinates to oxygen and not to sulfur, the complexes possible with Mg^{2+} are the Λ form of the B stereoisomer and the Δ form of the A stereoisomer. Yeast hexokinase exhibits marked stereospecificity for the B diastereoisomer of ATPβS, indicating that the active form of MgATP is as shown

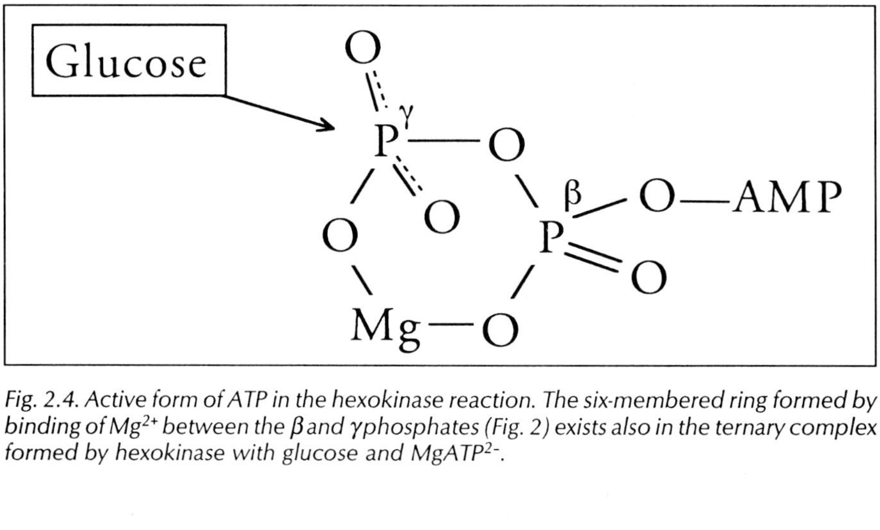

Fig. 2.4. Active form of ATP in the hexokinase reaction. The six-membered ring formed by binding of Mg^{2+} between the β and γ phosphates (Fig. 2) exists also in the ternary complex formed by hexokinase with glucose and $MgATP^{2-}$.

in Figure 2.6. Similar experiments carried out with liver hexokinase D and muscle hexokinase B have shown that in the presence of Mg^{2+} both enzymes exhibit a preference for the B diastereoisomer of ATPβS,[16] results qualitatively similar to those for yeast hexokinase B.[13,15] This information will be invaluable for understanding the detailed mechanism when the crystal structure of hexokinase D becomes available.

Kinetic studies also support this unity of mechanism. It is generally agreed, for example, that yeast hexokinase follows a random-order ternary-complex mechanism in which the preferred pathway is the addition of glucose before ATP and the release of glucose 6-phosphate after ADP,[17] and a similar pathway appeared to occur in wheat-germ hexokinase.[18] As we shall see in Chapter 6, flux-ratio studies of hexokinase D showed some degree of randomness, with a predominant ordered preferred pathway with glucose binding first and glucose 6-phosphate released last,[19] as in yeast and wheat germ. With the same technique, a similar order of addition of substrates and release of products has been found to predominate in hexokinase B.[20]

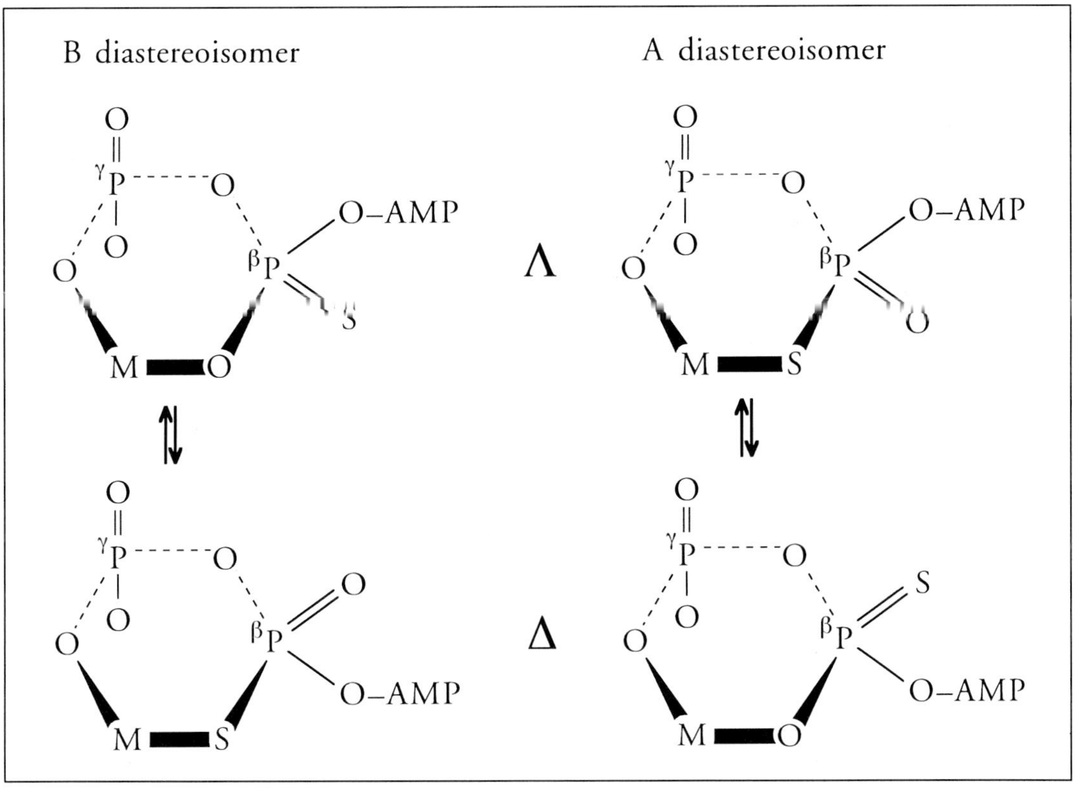

Fig. 2.5. Diastereoisomers of ATPβS. Each diastereoisomer has two screw senses designated Λ and Δ, but they do not necessarily exist for any given metal ion. For example, Mg^{2+} does not coordinate to sulfur, and so MgATPβS exists either as BΛ or as AΔ.

HOW HEXOKINASE D CAME TO BE CALLED GLUCOKINASE

The story of vertebrate hexokinase D could be said to have started in 1950 in the laboratory of Carl and Gerty Cori, who showed that under certain conditions liver extracts could phosphorylate glucose but not fructose; on this basis they postulated the existence of a glucokinase in the liver.[21] At the time their observation appeared very surprising, because it had been known ever since the first work on yeast hexokinase (ATP:D-hexose 6-phosphotransferase, EC 2.7.1.1) that this

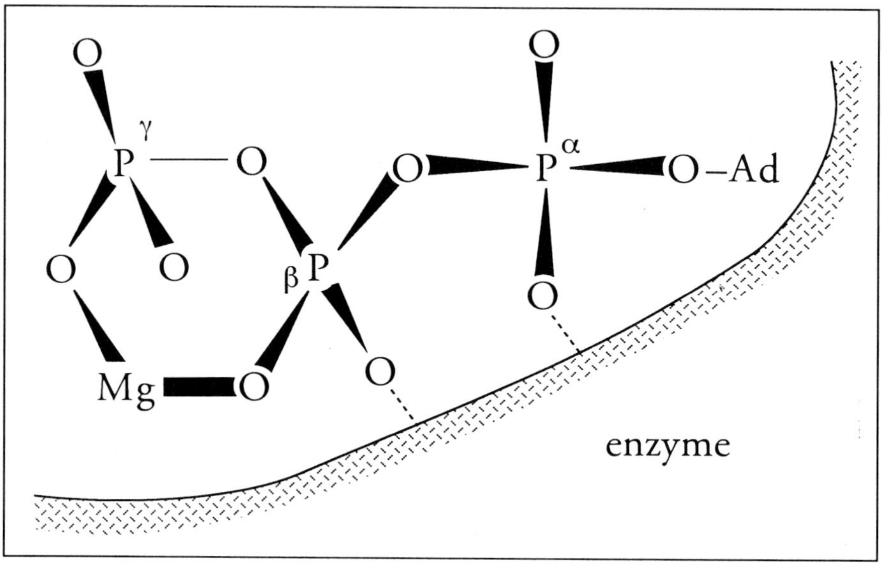

Fig. 2.6. Active form of $MgATP^{2-}$ in the hexokinase active site.

enzyme catalyzed the phosphorylation of the C6 position not only of glucose, but also of mannose and fructose.[5,21-23] In addition, mammalian hexokinases from brain and muscle exhibited the same broad sugar specificity.[5,24]

Some years later it was reported that in slices[25] and extracts[26] of rat liver the phosphorylation of glucose by the so-called glucokinase occurred with a high K_m. Consequently, Walker's discovery[27] of an enzyme in rat and guinea-pig liver that seemed to be rather specific for glucose, confirming the observations of the Coris' group, and with a rather low affinity for glucose, in agreement with those of Weinhouse's,[26] was of particular interest. This enzyme had a rather high half-saturation concentration for glucose, about 10 mM, and it coexisted in the liver with a hexokinase that more closely resembled the enzymes of brain and muscle, with a K_m of about 0.01 mM.[28,29] Sols's group[29] proposed that Cori's suggested name glucokinase (ATP:

D-glucose 6-phosphotransferase, EC 2.7.1.2) should be retained for the new enzyme, in part as recognition of the earlier observation in liver extracts.[21] (The reader must bear in mind the great intellectual influence of Carl Cori at that time, derived from his enormous contributions to the field of hexokinase and glucose metabolism). However, the imprecision of the data available at that time led the authors to add that "nevertheless a further characterization of the enzyme could make more appropriate some other name."

In summary, the newly discovered enzyme was characterized by a high half-saturation concentration of glucose (about 10 mM), a relatively low capacity for phosphorylating fructose at the concentration of 100 mM and an insensitivity to inhibition by glucose 6-phosphate at physiological concentrations. The other isoenzyme, which appeared to be a typical animal hexokinase as it had a very low K_m for glucose (0.01mM), phosphorylated fructose as well as glucose and was strongly inhibited by glucose-6-phosphate, was designated as hexokinase (EC 2.7.1.1).

Afterwards, when the two components of the rat-liver system were resolved into four isoenzymes, as we saw in the previous chapter, it was found that three of them shared the usual properties of hexokinases in relation to K_m for glucose and inhibition by glucose 6-phosphate, and they were identified as hexokinases A, B, and C according to their decreasing chromatographic mobility in DEAE-cellulose,[30,31] or hexokinases I, II, and III according to their increasing anodic mobility in electrophoresis.[32] The fourth isoenzyme, named hexokinase D or IV, continued to be called glucokinase because of the relative rates of phosphorylation of fructose and glucose with the purified isoenzymes at a sugar concentration of 100 mM; whereas the purified hexokinases A, B and C phosphorylated fructose 1.1 to 1.3 times faster than glucose, the phosphorylation ratio of fructose and glucose with hexokinase D was about 0.2.[31,33]

How specific is hexokinase D?

Since these first reports, therefore, there was general agreement that hexokinase D had a very low activity with fructose by comparison with the other animal hexokinases, an idea that derived mainly from the relative rates of phosphorylation of fructose and glucose at a single sugar concentration of 100 mM, as we have seen. A point to bear in mind, however, is that for hexokinases A, B, and C a concentration of 100 mM is saturating for both hexoses, whereas for hexokinase D it approaches saturation for glucose but it is well below saturation for fructose,[2] as can be seen in Table 2.1. A complication in the earlier studies was that the hexokinase D was not pure, and there was a feeling that at least part of the fructose-phosphorylating activity could be attributed to a contaminating enzyme.[31] Such contamination could perhaps explain an odd result from these workers, who reported a K_m

Table 2.1. Kinetic parameters for hexokinase isoenzymes with glucose and fructose as substrates at pH 7.5[1]

Isoenzyme	$K_{0.5}^{Glc}$ (mM)	$K_{0.5}^{Fru}$(mM)	$\frac{K_{0.5}^{Fru}}{K_{0.5}^{Glc}}$	$\frac{V^{Fru}}{V^{Glc}}$	$\frac{V^{Fru}/K_{0.5}^{Fru}}{V^{Glu}/K_{0.5}^{Glc}}$
Hexokinase A	0.044	3.1	70.5	1.1	0.016
Hexokinase B	0.130	3.0	23.1	1.2	0.052
Hexokinase C	0.020	1.2	60.0	1.3	0.022
Hexokinase D	7.5	420	56.0	2.4	0.043

1 Data of Cárdenas and coworkers.[2] *V* corresponds to the limiting velocity when the substrate concentration tends to infinity: it is normally known as the maximum velocity, $K_{0.5}$ is the half-saturation concentration, i.e. the substrate concentration at which the rate is 0.5*V*.

for fructose of 12 mM. Afterwards, when hexokinase D was purified it was clearly established that it could phosphorylate fructose, but with a much higher K_m.[2]

In spite of this consensus, however, numerous authors questioned whether "glucokinase" was a proper name for an enzyme with such a broad specificity;[1,6,33,34] Ureta et al.[1] (1979) in particular made an extensive analysis of the problem. Furthermore, very early Walker had pointed out that although the name glucokinase (and a different EC number) for this isoenzyme might be a useful guide to its physiological role it was unsatisfactory, as the enzyme was not wholly specific for glucose and could also catalyze the phosphorylation of mannose, fructose and 2-deoxyglucose, sugars that are also competitive inhibitors of the phosphorylation of glucose.[28] In these properties the liver enzyme differs from the glucokinases from *Aerobacter aerogenes*, *Brevibacterium fuscum*, *Bacillus stearothermophilus*, *Streptococcus mutans* OMZ 70, *Dictyostelium discoideum*, *Schistosoma mansoni*, and *Echinococcus granulosus*, which are truly specific for glucose; on the contrary, it resembles the multiple-substrate hexokinases of yeast and mammals (for specific references see articles 1 and 3). In particular, in *Streptococcus mutans*,[35] and also in *Aerobacter aerogenes*,[36] there are highly specific glucokinases that are not even inhibited by fructose or mannose, which is far from being the case for vertebrate hexokinase D.

Even if historically this relative capacity of phosphorylating fructose could have been a good argument for the denomination of glucokinase and a separate EC number this is no longer true. Assessing specificity by kinetic measurements at a single concentration, with substrates considered one at a time, was once quite standard. However, there is now widespread acceptance of Fersht's view that discrimination between substrates that are simultaneously present provides the only meaningful measure of specificity.[37] Thus, instead of comparing the relative rates at a single substrate concentration, which may appear to justify the idea that hexokinase D has some specificity, we must compare kinetic parameters that truly measure specificity according to Fersht's definition. A suitable measure is the ratio $(V_{Fru}/K_{0.5,Fru})/(V_{Glc}/K_{0.5,Glc})$, which has been used for many years in the hexokinase field under the name "phosphorylation coefficient," but more generally it is the ratio of the specificity constants for fructose and glucose.[38] The phosphorylation coefficient was introduced by Sols and Crane[24] and is a useful indicator of the relative susceptibility of a substrate to phosphorylation in relation to glucose.

For an enzyme obeying Michaelis-Menten kinetics the ratio k_{cat}/K_m is identical to $k_{cat}/K_{0.5}$, and is the best measure of specificity; although it is not strictly correct for a cooperative enzyme such as hexokinase D, it is preferable to other parameters. Table 2.1, taken from reference 2, shows that although hexokinase D has a much higher $K_{0.5}$ for fructose than for glucose, this is in no way a special property of hexokinase D because it is also true for hexokinases A, B, and C. On the other hand, hexokinase D has a much *higher* limiting rate with fructose than with glucose. (In fact, although the K_m for fructose is approximately 100 times higher than the half saturation constant for glucose, the limiting rate is more than double (see Fig. 2.7 and Tables 2.1 and 2.2). In the case of the other hexokinases the K_m is also about 100 times bigger for fructose than for glucose, but now the limiting rate is only slightly higher. This means that the specificity constant for fructose of the four isoenzymes is rather similar, and that for *all* of them, and not just for "glucokinase," glucose is a more specific substrate than fructose.

Table 2.2 shows the phosphorylation coefficients of hexokinase D with the different sugar substrates. I must emphasize that the parameters shown were obtained with purified enzyme of much higher specific activity and greater stability than had been used in the early work, before the cooperativity of the enzyme with respect to glucose and

Table 2.2. Phosphoryl acceptor specificity at pH 8.0[1]

Sugar	$K_{0.5}$ (mM)	h	$\frac{K_{0.5}^{Hexose}}{K_{0.5}^{Glucose}}$	$\frac{V^{Hexose}}{V^{Glucose}}$	$\frac{V^{Hexose}/K_{0.5}^{Hexose}}{V^{Glucose}/K_{0.5}^{Glucose}}$
Glucose	5	1.5	1	1	1
Mannose	8	1.5	1.6	0.9	0.56
2-Deoxyglucose	22	1.0	4.4	0.7	0.16
Fructose	325	1.0	65	2.6	0.04
Fructofuranose	65	1.0	16.2	2.6	0.16

1. Data of Cárdenas and coworkers.[2,61] The parameters are as in Table 2.1 apart from *h*, the Hill coefficient, which is an empirical measurement of the degree of cooperativity: if *h* = 1.0 there is no cooperativity, if *h* is less than 1.0 there is negative cooperativity, and if it is greater than 1.0 there is positive cooperativity.

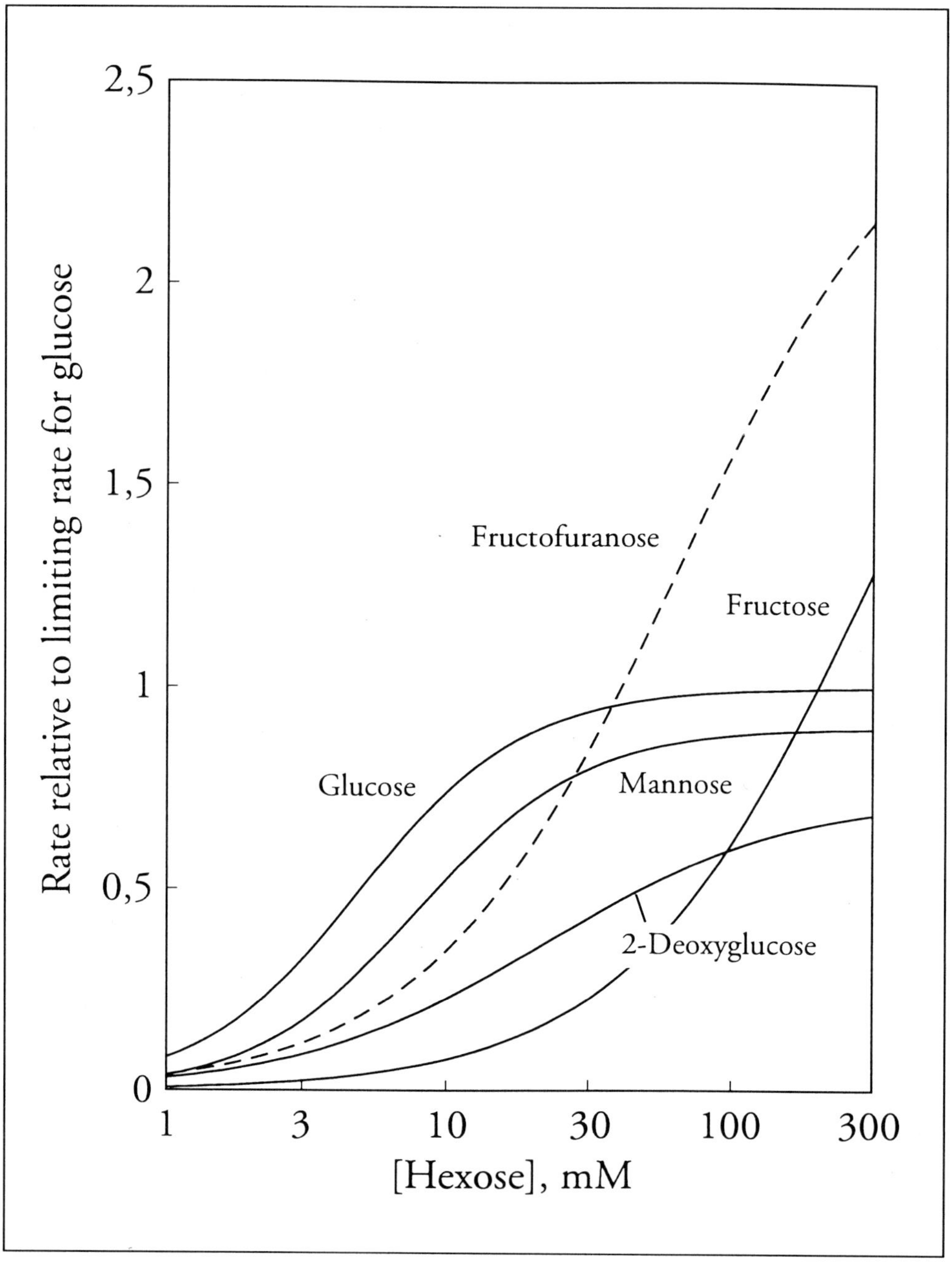

Fig. 2.7. Rates of phosphorylation of various sugars by hexokinase D. The curves were calculated from the parameter values listed in Table 2.2. Note that fructose is phosphorylated faster than any other sugar at high concentrations, or even at moderate concentrations if one recognizes that the reactive tautomer fructofuranose represents only about 25% of the fructose in aqueous solution. Reproduced with permission from Arch Biol Med Exp 1985; 18:273-284.

mannose was known.[31,39,40] It is quite wrong and misleading, therefore, to continue quoting the kinetic parameters from these first studies. For instance, the relative maximum rate of hexokinase D with fructose is not 8%, as claimed in a very recent review,[41] but 260%, an error of more than a factor of 30! The difference in pH affects the values very slightly, as will be seen in the next chapter. If it is taken into account that only the furanose tautomer of fructose is a substrate for the hexokinases,[21] and that in aqueous solution at 30°C only about 25% of fructose is in the furanose form,[42] a phosphorylation coefficient for this tautomeric form four times higher than that for fructose can be calculated. This value is then similar to that for 2-deoxyglucose, or even higher if the nonsubstrate tautomer acts as a competitive inhibitor. (As nearly all textbooks illustrate the furanose structure of fructose, it is perhaps important to emphasize that it is the six-membered pyranose ring that predominates in aqueous solution under ordinary conditions).

The similarity in the phosphorylation coefficient for fructose of all the hexokinases (Table 2.1) makes it clear that the name "glucokinase" for hexokinase D is a misnomer. In addition it appears there is a unity of chemical mechanism between hexokinase D and the other hexokinases,[16,34] as we mentioned above.

Something to consider, however, is that in view of the concentrations of fructose attainable inside the hepatocytes (the concentration found in the portal blood after a fructose-rich meal being only about 2 mM), "glucokinase" in vivo would never phosphorylate fructose to a significant extent. Glucose is therefore the only sugar that it phosphorylates under physiological conditions, and this fact is still sometimes given as a justification for the name "glucokinase." In my opinion this argument is misleading, because although it is true in principle that the other isoenzymes (the low-K_m hexokinases) *could* phosphorylate fructose in vivo, it is not at all probable that they do so in fact, as the metabolism of fructose in the liver occurs mainly by phosphorylation of fructose at C1 through the action of a ketohexokinase, also called fructokinase,[43] which has a much lower K_m for fructose than any of the hexokinases. Consequently, it seems that neither hexokinase D nor the other hexokinases are responsible for the phosphorylation of fructose in vivo. If the in vivo argument is applied, therefore, they would *all* be "glucokinases."

Instead of glucokinase, therefore, the name hexokinase D (or hexokinase type IV) within the classification number of EC 2.7.1.1 is more appropriate, and the name glucokinase should be reserved for those enzymes known to be specific for glucose, such as those isolated from several invertebrates and microorganisms, as mentioned above. As a matter of fact, the Nomenclature Committee of the International Union of Biochemistry and Molecular Biology has accepted this argument and since 1984 vertebrate "glucokinase" has been listed in

Enzyme Nomenclature under the same EC number as the other isoenzymes, and the recommended name is hexokinase D or hexokinase type IV. Unfortunately, in spite of this recommendation there are still publications that attribute to hexokinase D the wrong EC number, a serious problem as readers can be misled into thinking that vertebrate hexokinase D is a real glucokinase. It is especially worrying to see this in articles[44] that present information of the greatest importance.

If one considers that in vivo all the mammalian hexokinases phosphorylate mainly glucose, that they all have the highest phosphorylation coefficient with glucose, whereas they all have a similarly low phosphorylation coefficient with fructose and they appear to follow the same chemical mechanism, it appears arbitrary to distinguish hexokinase D (or IV) as glucokinase. It is true, of course, that hexokinase D differs from the other three isoenzymes in several respects, such as molecular mass, kinetic behavior, specific gene expression and inducibility in liver, but similar variations are common for isoenzymes. For example, pyruvate kinase shows important kinetic differences between type L (liver) and type M (muscle), and liver contains two isoenzymes: isoenzyme L, like hexokinase D, is confined to hepatocytes, whereas isoenzyme K is present in nonparenchymal cells. Isoenzyme L depends on the presence of glucose and insulin, whereas isoenzyme K is constitutive. However, no one has thought it would be helpful to distinguish between them by means of names that falsely suggest that the difference is a difference of specificity, such as calling one of them pyruvate kinase and the other oxoacid kinase.

Apart from the low affinity for glucose, other properties often regarded as special features of hexokinase D are its much lower inhibition constant (around 60 mM)[45] for inhibition by glucose 6-phosphate compared with those in the micromolar range shown by hexokinases A, B and C, and its lower molecular mass (about 50 kDa compared with about 100 kDa).[33] These supposedly special properties require some comments. First, the relative insensitivity to glucose 6-phosphate is not unique to hexokinase D, as it applies also to hexokinase C, which despite having the highest affinity of all for glucose (see Table 2.1) has an inhibition constant for glucose 6-phosphate at least an order of magnitude higher than those for hexokinases A and B.[46] So, comparing $K_{0.5}^{Glucose}/K_i^{Glc6P}$ values for isoenzymes C and D we have 0.02/0.18 for hexokinase C and 5/60 for hexokinase D, virtually the same value of 0.1 for the two. (Of course, the inhibition of hexokinase D by glucose 6-phosphate must be physiologically irrelevant).

Until fairly recently the strong inhibition of hexokinase A by glucose 6-phosphate was thought to be due to interaction of this product at an allosteric site situated in the *N*-terminal half of the molecule,[47,48] whereas the *C*-terminal half was responsible for all the catalytic activity.[49,50] As there is good evidence for gene duplication in vertebrate hexokinases, several authors have suggested that the site for glucose

6-phosphate evolved from a duplicated active site (see the review by Ureta and coworkers[51] and references therein). However, although the *N*-terminal half of hexokinase A can indeed bind glucose 6-phosphate, it is not responsible for the inhibition, though it could play a role in the control of the binding of hexokinase A to mitochondria.[52] Findings from different groups, using different approaches, support the view that the *C*-terminal half, which shows 53% identity with hexokinase D,[53] and has a similar molecular mass, provides both activities, not only catalysis but also high sensitivity to inhibition by glucose 6-phosphate.[52,54,55] All of these considerations make it reasonable to suppose that all enzymes catalyzing the phosphorylation of hexoses follow essentially the same mechanism, and that the range of types of kinetic behavior derive from quantitative differences in rate constants.

It is true that the physiological importance of hexokinase D goes beyond its status as the predominant hexokinase in the liver, as it is also by far the main hexokinase in the hepatocytes. However, its importance does not justify a special name, because hexokinases A, B and D are all important in one way or another. Glucose is the main oxidizable substrate used by the majority of mammalian tissues as their source of energy for endergonic biosynthetic processes and other forms of work. Many tissues can use other metabolites such as fatty acids as an energy source, but some, such as brain, seem to be absolutely dependent upon glucose: in brain the main hexokinase is hexokinase A, and so for the physiology of the organism hexokinase A is no less important than hexokinase D; its importance is just different. Hexokinase B likewise fulfils the particular needs of the mammalian muscle.

I do not discuss here the adaptive character of liver hexokinase D, because that is not related to specificity and should not be given as an argument for the name glucokinase. As we shall see in Chapter 3, the pancreatic hexokinase D, which has also been called glucokinase, is mainly constitutive.

Hers has argued that "there is no reason to give up the easily remembered name 'glucokinase' in favour of 'hexokinase IV,' a numerical classification that very few people would be able to justify and that can lead to confusion as, for instance, in calling the enzyme 'hexokinase VI'".[56] The name "hexokinase IV" mentioned here refers to hexokinase type IV, one of the names listed by IUBMB in *Enzyme Nomenclature*, but "hexokinase VI" just reproduces what was almost certainly a typographical error in the letter of Iynedjian and Girard[57] referred to. There was nothing in the rest of their letter to suggest that they were confused or that they were proposing "hexokinase VI" as a name; on the contrary, their main point was to emphasize the confusion that results from misclassifying the enzyme with the specific glucokinases of EC 2.7.1.2. Hers's argument that an archaic name that misleads readers about the specificity is "easily remembered" was doubtless once advanced in favor of retaining such names as heterophosphatase and carboxydismutase (for hexokinase and ribulose-bisphosphate carboxylase respectively), but all such names have been eliminated from *Enzyme Nomenclature* and few would seriously argue for their revival.

The name hexokinase D, in any case, does not have the problem of a numerical classification and it is easily remembered. Although the original letter D was given because it was the fourth hexokinase to elute from the chromatographic column,[30] D is not only the fourth letter of the alphabet but is also the initial letter of diet and diabetes, both relevant to the level of hexokinase D, as we saw in the previous chapter. So it should not be difficult to remember that liver hexokinase D is the **diet-dependent** hexokinase, with a level that **decreases** in **diabetes.**

HOW MUCH DOES THE NAME MATTER?

This point of nomenclature—hexokinase or "glucokinase"—is more serious than just a semantic discussion, as some people may believe, because researchers outside the field but interested in related topics, such as protein evolution[58,59] or diabetes,[62] or even researchers interested in hexokinase D[63] only in a peripheral way, are misled by the name, and they make the natural assumption that vertebrate "glucokinase" is similar to the real glucokinase of bacteria, which is highly specific.[35,36] Griffin and colleagues,[58] for example, appear surprised at finding that rat "glucokinase" was most closely related to the *C*-terminal halves of the hexokinases, whereas yeast glucokinase was most closely related to the yeast hexokinases. They say "this observation suggests that the duplication leading to rat glucokinase was a separate (and more recent) event from the duplication leading to the yeast glucokinase, implying that the glucokinase activities arose twice in evolution." In reality, of course, there is no need to postulate that glucokinase activity arose twice in evolution because "vertebrate glucokinase" is not a glucokinase, and nor is "yeast glucokinase"!

References

1. Ureta T, Radojkovic J, Lagos R et al. Phylogenetic and ontogenetic studies of glucose phosphorylating isoenzymes of vertebrates. Arch Biol Med Exp 1979; 12:587-609.
2. Cárdenas ML, Rabajille E, Niemeyer H. Fructose: a good substrate for rat-liver "glucokinase" (hexokinase D). Biochem J 1984; 222:363-370.
3. Cornish-Bowden A, Cárdenas ML. Hexokinase and "glucokinase" in liver metabolism. Trends Biochem Sci 1991; 16:281-282. Additional correspondence in vol. 17, p. 59.
4. Meyerhof O. Über die enzymatische Milchsäurebildung im Muskelextrakt. III. Mitteilung: Die Milchsäurebildung aus den gärfähigen Hexosen. Biochem Z 1927; 183:176-215.
5. Crane RK. Hexokinases and pentokinases. In: Boyer PD, Lardy H,

Myrbäck K, eds. The Enzymes, 2nd ed. New York: Academic Press, 1962; 6:47-66.

6. Walker DG. The nature and function of hexokinases in animal tissues. Essays Biochem 1966; 2:33-37.
7. Pollard-Knight D, Cornish-Bowden A. Kinetics of hexokinase D ("glucokinase") with inosine triphosphate as phosphate donor. Biochem J 1987; 245:625-629.
8. Cohn M. Some mechanisms of cleavage of adenosine triphosphate and 1,3-diphosphoglyceric acid. Biochim Biophys Acta 1956; 20:92-99.
9. Cohn M, Hughes TR. Nuclear magnetic resonance spectra of adenosine di- and triphosphate. II Effect of complexing with divalent metal ions. J Biol Chem 1962; 237:176-181.
10. Lowe G, Potter BVL. The stereochemical course of yeast hexokinase-catalysed phosphoryl transfer by using adenosine 5'-[$\gamma(S)$-^{16}O,^{17}O,^{18}O] triphosphate as substrate. Biochem J 1981; 199:227-233.
11. Rose IA. Mechanism of phosphoryl transfer by hexokinase. Biochem Biophys Res Comm 1980; 94:573-578.
12. Spector LB. In: Covalent catalysis by enzymes. New York: Springer-Verlag, 1982.
13. Cornelius RD, Cleland WW. Substrate activity of (adenosine triphosphato) tetraamminecobalt(III) with yeast hexokinase and separation of diastereomers using the enzyme. Biochemistry 1978; 17:3279-3286.
14. Pollard-Knight D, Potter BVL, Cullis PM et al. The stereochemical course of phosphoryl transfer catalysed by glucokinase. Biochem J 1982; 201:421-423.
15. Jaffe EK, Cohn M. Diastereomers of the nucleoside phosphorothioates as probes of the structure of the metal nucleotide substrates and of the nucleotide binding site of yeast hexokinase. J Biol Chem 1979; 254:10839-10845.
16. Darby M, Trayer IP. Metal-nucleotide structure at the active sites of the mammalian hexokinases. Eur J Biochem 1983; 129:555-560.
17. Colowick SP. The hexokinases. In: Boyer PD, ed. The Enzymes. 3rd ed. New York: Academic Press, 1973; 9:1-48.
18. Meunier J-C, Buc J, Navarro A. et al. Regulatory behavior of monomeric enzymes. 2. A wheat-germ hexokinase as a mnemonical enzyme. Eur J Biochem 1974; 49:209-223.
19. Gregoriou M, Trayer IP, Cornish-Bowden A. Isotope exchange evidence for an ordered mechanism for rat-liver glucokinase, a monomeric co-operative enzyme. Biochemistry 1981; 20:499-506.
20. Gregoriou M, Trayer IP, Cornish-Bowden A. Isotope-exchange evidence that glucose 6-phosphate inhibits rat-muscle hexokinase II at an allosteric site. Eur J Biochem 1983; 134:283-288.
21. Slein MW, Cori GT, Cori CF. A comparative study of hexokinase from yeast and animal tissues. J Biol Chem 1950; 186:763-780.
22. Colowick SP, Kalckar HM. The role of myokinase in transphosphorylations. 1. The enzymatic phosphorylation of hexoses by adenyl

pyrophosphate. J Biol Chem 1943; 148:117-126.
23. Berger L, Slein MW, Colowick SP et al. Isolation of hexokinase from baker's yeast. J Gen Physiol 1946; 29:379-391.
24. Sols A, Crane RK. Substrate specificity of brain hexokinase J Biol Chem 1954; 210:581-595.
25. Cahill GF Jr, Hastings AB, Ashmore J et al. Studies on carbohydrate metabolism in rat liver slices. X. factors in the regulation of pathways of glucose metabolism. J Biol Chem 1958; 230:125-135.
26. DiPietro DL, Sharma C, Weinhouse S. Studies on glucose phosphorylation in rat liver. Biochemistry 1962; 1:455-462.
27. Walker DG. The development of hepatic hexokinases after birth. Biochem J 1962; 84:118P-119P.
28. Walker DG. On the presence of two soluble glucose-phosphorylating enzymes in adult liver and the development of one of these after birth. Biochim Biophys Acta 1963; 77:209-266.
29. Viñuela E, Salas M, Sols A. Glucokinase and hexokinase in liver in relation to glycogen synthesis. J Biol Chem 1963; 238:PC1175-1177.
30. González C, Ureta T, Sánchez R et al. Multiple molecular forms of ATP:hexose 6-phosphotransferase from rat liver. Biochem Biophys Res Commun 1964; 16:347-352.
31. González C, Ureta T, Babul J et al. Characterization of isoenzymes of adenosine triphosphate:D-hexose 6-phosphotransferase from rat liver. Biochemistry 1967; 6:460-468.
32. Katzen HM, Soderman DD, Nitowski HM. Kinetic and electrophoretic evidence for multiple forms of glucose-ATP phosphotransferase activity from human cell cultures and rat liver.Biochem Biophys Res Commun 1965; 19:377-382.
33. Niemeyer H, Ureta T, Clark-Turri L. Adaptive character of liver glucokinase. Mol Cell Biochem 1975; 6:109-126.
34. Pollard-Knight D, Cornish-Bowden A. Mechanism of liver glucokinase. Mol Cell Biochem 1982; 44:71-80.
35. Porter EV, Chassy BM, Holmlund CE. Purification and kinetic characterization of a specific glucokinase from *Streptococcus mutans* OMZ70 cells. Biochim Biophys Acta 1982; 709:178-186.
36. Kamel MY, Allison DP, Anderson RL. Stereospecific D-glucokinase of *Aerobacter aerogenes*. Purification and properties. J Biol Chem 1966; 241:690-694.
37. Fersht A. In: Enzyme Structure and Mechanism. 2nd ed. New York: WH Freeman, 1985:347-368.
38. International Union of Biochemistry. Nomenclature Committee. Symbolism and terminology in enzyme kinetics. Recommendations 1981. Eur J Biochem 1982; 128:281-291.
39. Salas J, Viñuela E, Salas M et al. Glucokinase of rabbit liver. Purification and properties. J Biol Chem 1965; 240:1014-1018.
40. Parry MJ, Walker DG. Purification and properties of adenosine 5'-triphosphate-D-glucose 6-phosphotransferase from rat liver. Biochem J 1966;

99:266-274.

41. Randle PJ. Glucokinase and candidate genes for type 2 (non-insulin-dependent) diabetes mellitus. Diabetologia 1993; 36:269-275.
42. Que L, Gray GR. ^{13}C-nuclear magnetic resonance spectra and the tautomeric equilibria of ketohexoses in solution. Biochemistry 1974; 13:146-153.
43. Hers HG. The conversion of fructose-1-C^{14} and sorbitol-1-C^{14} to liver and muscle glycogen in the rat. J Biol Chem 1955; 214:373-381.
44. Efrat S, Leiser M, Wu Y-J et al. Ribozyme-mediated attenuation of pancreatic β-cell glucokinase expression in transgenic mice results in impaired glucose-induced insulin secretion. Proc Natl Acad Sci USA 1994; 91:2051-2055.
45. Storer AC, Cornish-Bowden A. Kinetic evidence for a "mnemonical" mechanism for rat liver glucokinase. Biochem J 1977; 165:61-69.
46. Radojkovic J, Ureta T. Hexokinase isoenzymes from the Novikoff hepatoma. Purification, kinetic and structural characterization, with emphasis on hexokinase C. Biochem J 1987; 242:895-903.
47. White TK, Wilson JE. Rat brain hexokinase: location of the allosteric regulatory site in a structural domain at the *N*-terminus of the enzyme. Arch Biochem Biophys 1987; 259:402-411.
48. Schwab DA, Wilson JE. Complete amino acid sequence of rat brain hexokinase, deduced from the cloned cDNA, and proposed structure of a mammalian hexokinase. Proc Natl Acad Sci USA 1989; 86:2563-2567.
49. Nemat-Gorgani M, Wilson JE. Rat brain hexokinase: location of the substrate-binding site in a structural domain at the C-terminus of the enzyme. Arch Biochem Biophys 1986; 251:97-103.
50. Schirch DM, Wilson JE. Rat brain hexokinase: location of the substrate binding site in a structural domain at the *C*-terminus of the enzyme. Arch Biochem Biophys 1987; 254:385-396.
51. Ureta T, Medina C, Preller A. The evolution of hexokinases. Arch Biol Med Exp 1987; 20:343-357.
52. Arora KK, Filburn CR, Pedersen PL. Structure/function relationships in hexokinase. Site-directed mutational analyses and characterization of overexpressed fragments implicate different functions for the *N*- and *C*-terminal halves of the enzyme. J Biol Chem 1993; 268:18259-18266.
53. Fothergill-Gilmore LA, Michels PAM. Evolution of glycolysis. Prog Biophys Mol Biol 1993; 59:105-235.
54. White TK, Wilson JE. Isolation and characterization of the discrete *N*- and *C*-terminal halves of rat brain hexokinase: retention of full catalytic activity in the isolated *C*-terminal half. Arch Biochem Biophys 1989; 174:375-393.
55. Magnani M, Bianchi M, Casabianca A et al. A recombinant human "mini"-hexokinase is catalytically active and regulated by hexose 6-phosphates. Biochem J 1992; 285:193-199.
56. Hers HG. Regulation of glucokinase and its role in liver metabolism. Trends Biochem Sci 1992; 17:59.

57. Iynedjian PB, Girard J. Nomenclature for mammalian glucokinase. Biochem J 1991; 275:821.
58. Griffin LD, Gelb BD, Wheeler DA et al. Mammalian hexokinase 1: evolutionary conservation and structure to function analysis. Genomics 1991; 11:1014-1024.
59. Bork P, Sander C, Valencia A. Convergent evolution of similar enzymatic function on different protein folds: the hexokinase, ribokinase, and galactokinase families of sugar kinases. Protein Science 1993; 2:31-40.
60. Cárdenas ML, Rabajille E Trayer IP et al. Cooperative interactions in hexokinase D ("glucokinase"). Kinetic and fluorescence studies. Arch Biol Med Exp 1985; 18:273-284.
61. Cárdenas ML, Rabajille E, Niemeyer, H. Suppression of kinetic cooperativity of hexokinase D ("glucokinase") by competitive inhibitors. Eur J Biochem 1984; 145:163-171.
62. Unger RH. Diabetic hyperglycemia: Link to impaired glucose transport in pancreatic β cells. Science 1991; 251:1200-1205.
63. St. Charles R, Harrison RW, Bell GI et al. Molecular model of human β-cell glucokinase built by analogy to the crystal structure of yeast hexokinase B. Diabetes 1994; 43:784-791.

CHAPTER 3

Kinetic and Structural Properties of Hepatic Hexokinase D, a Monomeric Cooperative Enzyme

The kinetic properties of hexokinase D were well established more than a decade ago. However, they appear to have been partially forgotten, and now that it is desirable to compare the kinetic properties of the many mutants that have become available, especially in relation to NIDDM (see Chapter 5) it is important to know what these properties are, and to know them in sufficient detail to understand the precautions needed for avoiding artifacts. As mutant enzymes are often less stable than the natural forms it is crucial to know which stabilizing conditions affect the cooperativity and which do not. For reviews the reader is referred especially to references 1-3.

COOPERATIVE KINETICS OF HEXOKINASE D WITH GLUCOSE

Although the first studies of rat liver hexokinase D suggested a hyperbolic saturation function with respect to glucose,[4-8] and likewise for the rabbit enzyme,[9] some anomalies in the reported graphs raised doubts about the reality of the Michaelian kinetics. For instance, in the work of González and coworkers there was a "flattening" of the saturation curve at high substrate concentrations that was interpreted as inhibition by excess of substrate.[4] (Although at low substrate concentrations the curve for substrate inhibition is quite different from a sigmoid curve, the two curves are quite similar at high concentrations, especially if the substrate inhibition is sufficiently mild for no maximum to be visible). Moreover, Parry and Walker had described a biphasic curve in a double reciprocal plot, which they interpreted as due to fragments of enzyme with modified catalytic properties that appeared as artifacts of the purification and had no physiological meaning.[6] In reality their points suggested upward curvature, and the same could be seen in the work of Pilkis.[7] This is what one should expect if there is positive cooperativity; the existence of more than one form of enzyme, as suggested by Parry and Walker,[6] would have produced downward curvature, mimicking *negative* cooperativity. In other words the signs of positive cooperativity with respect to glucose were present in the original data of various groups, but were not immediately recognized as such.

It was only some years later that Niemeyer's group realized that the enzymes of hamster and rat displayed a characteristic that had passed unnoticed before and which could explain the observed anomalies; the saturation function for glucose was sigmoid, with a Hill coefficient of 1.5.[10,11] The cooperativity with glucose is now well established, and our results were soon confirmed by Storer and Cornish-Bowden,[12,13] and later by others.[14-16]

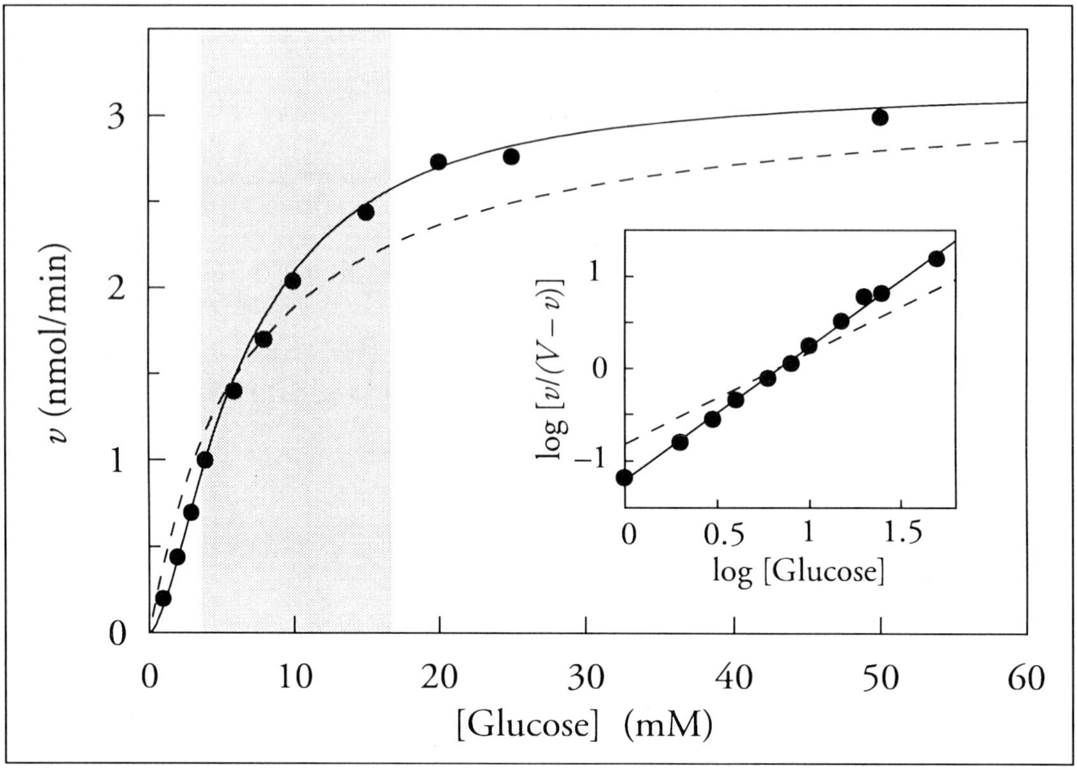

Fig. 3.1. Saturation function for glucose of rat liver hexokinase D. Partially purified enzyme was assayed at pH 7.5 and 30°C in the presence of 5 mM dithiothreitol and 100 mM K^+. The glucose 6-phosphate formed was measured by means of the coupled assay with glucose 6-phosphate dehydrogenase. The h and $K_{0.5}$ values were 1.58 and 6.6 mM respectively, and were obtained from the Hill plot shown in the inset. The broken line represents the theoretical curve for an enzyme exhibiting normal Michaelis-Menten kinetics (h = 1) with the same values of V_{max} and $K_{0.5}$ as the experimental curve. The shaded region shows the estimated range of variation of blood glucose concentration under different conditions of carbohydrate ingestion. Adapted from Niemeyer H, Cárdenas ML, Rabajille E, Ureta T, Clark-Turri, L, Peñaranda J. Enzyme 1975; 20:321-333.

This kinetic cooperativity was of interest because it could contribute to the regulation of the uptake of glucose in the hepatocyte, as will be discussed in Chapter 6. It is a distinctive characteristic of hexokinase D, as the low-K_m hexokinases display either hyperbolic kinetics (hexokinases A and B) or inhibition by excess of substrate (hexokinase C).[10,17]

A typical result obtained at pH 7.5 with partially purified rat liver hexokinase D is illustrated in Figure 3.1. This shows a sigmoidal saturation function for glucose with a Hill coefficient of 1.6 and a half-saturation concentration of about 7.0 mM.[11] The Hill coefficient is an empirical measure of the degree of cooperativity without specific mechanistic implications, and for this reason it is very convenient. The shaded area in Figure 3.1 represents the range of variation of glucose concentration in the portal vein under different conditions of carbohydrate ingestion. As the degree of cooperativity is small (h = 1.5) the sigmoid curve illustrated in Figure 3.1 is not obviously S-shaped, as it would be if the Hill coefficient were greater than 2.0. Despite the fact that the sigmoidicity is slight, the saturation curve of hexokinase D with glucose is not well fitted by the Michaelis-Menten equation, as will be analyzed in Figure 3.3. The deviations from Michaelis-Menten behavior become much more evident if one examines the results with the usual linear transformations of the Michaelis-Menten equation, as shown in Figure 3.2. Even this small degree of cooperativity is physiologically important, as discussed in Chapter 6 (see Fig. 6.1). Something to bear in mind is that in a semilogarithmic plot like that in Figure 6.1 *any* saturation curve looks S-shaped, so in that plot the corresponding curves for hexokinases A and B are S-shaped despite the lack of cooperativity with respect to glucose.

Storer and Cornish-Bowden reported a similar Hill coefficient (1.5-1.6), but a lower $K_{0.5}$ of 5 mM glucose;[12] this can be explained by the pH difference, as they worked at pH 8, where the $K_{0.5}$ is slightly lower,[18] as we shall see shortly. Both of these values for the half-saturation concentration for glucose are somewhat lower than the values of 10-20 mM reported by earlier workers.[6-9] The discrepancy can be explained by the fact that the curvature of the "linear" plots was overlooked or its importance underestimated, and also because their enzyme may have been partially inactivated, as discussed below.

As described in Chapter 5, pancreas contains an enzyme similar to hepatic hexokinase D, coded by the same gene even though it is not exactly the same protein. It shows the same cooperativity with respect to glucose as the hepatic enzyme, and its kinetic parameters are likewise similar.[19]

At the time when we described the kinetic cooperativity of hepatic hexokinase D with respect to glucose,[11] its quaternary structure was still unknown and we were consequently unable to assign a particular mechanism to the sigmoidal behavior but suggested instead several

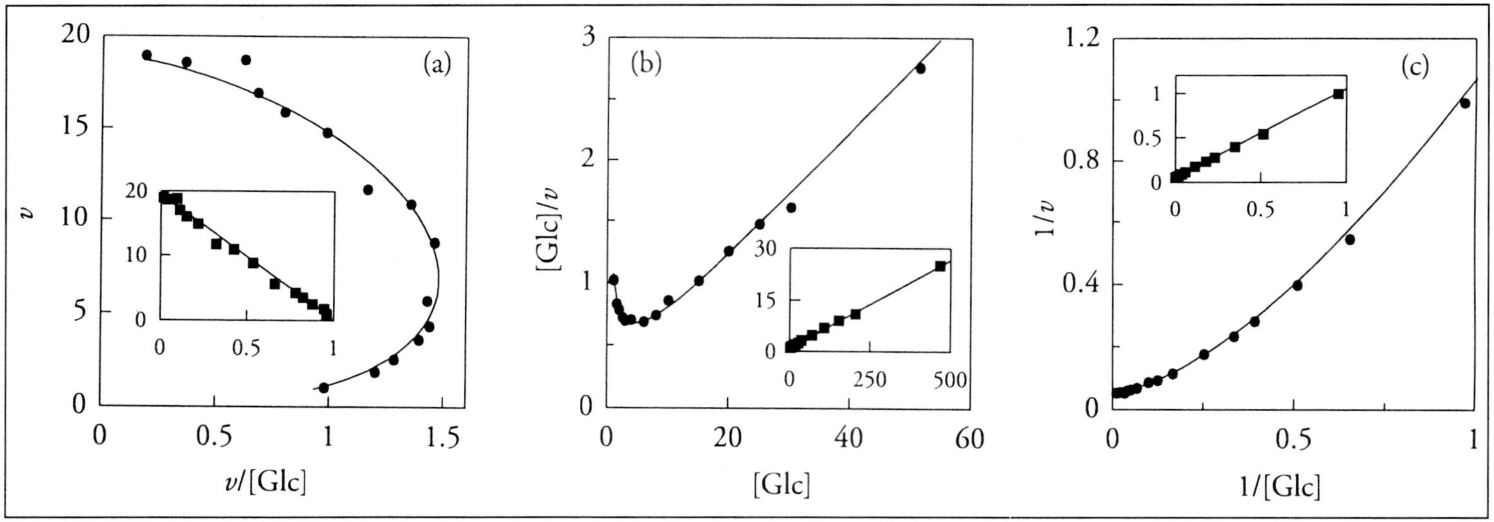

Fig. 3.2. Transformed plots of the saturation function for glucose of rat liver hexokinase D: (a) Eadie-Hofstee, (b) Hanes-Woolf, (c) Lineweaver-Burk. A similar experiment to that shown in Fig. 3.1 is illustrated, done with a higher concentration of enzyme and at a different stage of purification. The kinetic parameters were $h = 1.56$, $K_{0.5} = 6.7$ mM. The insets show the same graphs but with the glucose concentration raised to the power h. In the case of the Lineweaver-Burk and Hanes-Woolf transformations some of the points at the highest glucose concentrations have been omitted from the insets in the interests of clarity. Reproduced from the author's doctoral thesis.[58]

possibilities. We suspected that it was a monomer on the basis of its molecular mass and of the information that existed on the quaternary structures and molecular masses of the hexokinases of brain, wheat germ and yeast, but in any case the quaternary structure of hexokinase D under assay conditions was completely unknown. These questions were resolved in the years that followed,[20,21] when hexokinase D was clearly established to be monomeric under all conditions. Its monomeric structure has thus been an objective certainty since the end of the 1970s, and not merely a "consensus," as some recent authors seem to suppose.[22]

The degree of cooperativity does not depend on the method used for determining the enzyme activity,[11] and similar Hill coefficients are obtained whether the activity is determined spectrophotometrically by coupling the production of either glucose 6-phosphate or ADP to other reactions, or whether one avoids the use of coupling enzymes by directly measuring the [^{14}C]glucose 6-phosphate formed from [^{14}C]glucose,[23] or by measuring the release of protons with a pHstat.[12]

As the kinetic characteristics of an enzyme can be influenced by the presence of other proteins that interact with it, the kinetic behavior can vary with the degree of purification. We accordingly studied whether the cooperativity of hexokinase D varied with its purity after it was separated from the other isoenzymes, and we found no variation; this agreed with the report of Storer and Cornish-Bowden that homogeneous and partially purified preparations differ only in specific activity.[12] It is now known that a protein does exist that affects the kinetic parameters of hexokinase D, but it becomes separated from hexokinase D when this isoenzyme is separated from the others, and consequently neither our group nor those of Walker and Cornish-Bowden were able to detect it. It will be discussed in Chapter 7.

Cooperative phosphorylation of glucose in situ

The cooperativity is also observed both in hepatocyte extracts and in situ in isolated hepatocytes,[24] the formation of glucose 6-phosphate in the intact cell being estimated radiochemically using 2-^{3}H-glucose, assaying the $^{3}H_2O$ formed as described by Katz and coworkers.[25,26] In the experiments with extracts of isolated hepatocytes, the saturation curve was sigmoidal with a Hill coefficient of 1.5 and a $K_{0.5}$ of 9 mM,[24] similar to what is obtained with the purified enzyme. Similarly, glucose phosphorylation in situ in isolated hepatocytes is also sigmoidal, with a Hill coefficient of 1.8 and a $K_{0.5}$ of 20 mM glucose.[24] This agreed with the sigmoidal saturation curve for glucose utilization obtained by Katz's group some time earlier,[26] but the much higher half-saturation concentration of glucose[24,26] appeared surprising at the time; this can now probably be explained by the existence of the regulatory protein discussed in Chapter 7.

These results with isolated hepatocytes were very important, because they gave support to the physiological significance of the sigmoidal kinetics of isolated hexokinase D with respect to glucose. On the other hand the discovery of the kinetic cooperativity of the purified enzyme was also of great importance because it allowed the sigmoidicity of glucose utilization in hepatocytes[24,26] to be explained on the basis of the kinetic properties of hexokinase D without the need to invoke other hypotheses such as the action of two opposing processes, for example phosphorylation by hexokinase D and dephosphorylation by glucose 6-phosphatase, as in the original proposal of Katz's group.[26]

Cycling in liver between glucose and glucose-6-phosphate

The physiological advantage of the kinetic cooperativity of hexokinase D with respect to glucose is more or less obvious, as it makes the phosphorylation of glucose in the liver more sensitive to variations in the blood glucose concentration. However, it is less clear how this kinetic cooperativity affects the cycling between glucose and glucose-6-phosphate, for example in the analyses made by Newsholme and Start[27] and by Hue and Hers.[28] Two important effects not taken into account by those analyses were the sigmoidal dependence of the rate on the glucose concentration and the effect of MgATP concentration on both the rate and the sigmoidicity (see below), neither of which had been reported at that time. Unfortunately, even recent studies of hepatic glucose fluxes often fail to take account of the sigmoidicity. For example Barzilai and Rossetti argued that changes in glucose-6-phosphatase and hexokinase D activities could account for only a small portion of the in vivo inhibition of hepatic glucose flux by insulin,[29] implying the need for additional mechanisms for short-term regulation of hepatic glucose fluxes. However, as their studies took no account either of the sigmoidicity or of the regulatory protein, their conclusions cannot be regarded as established. A further complication is that the two enzymes are not uniformly distributed in the liver, hexokinase D predominating in perivenous and glucose 6-phosphatase in periportal hepatocytes: this will be discussed in Chapter 4.

Estimation of kinetic parameters: kinetic studies in hexokinase D mutants

Figures 3.1 and 3.2 illustrate two important concepts related to the kinetic cooperativity of hexokinase D with respect to glucose. Both need to be borne in mind in kinetic studies, especially now that the gene for hexokinase D has been cloned and can readily be expressed in *Escherichia coli*, allowing the production of various kinds of mutants by site-directed mutagenesis for structure-function studies.[22] Examination of the Hill plot in Figure 3.1 shows that it is linear over a relatively wide range of glucose concentrations above and below the half-saturation concentration, indicating that the cooperativity is essentially constant over this range. This idea is also illustrated in Figure 3.2, where the double-reciprocal plot is parabolic (to a first approximation), with no genuine linear segments. It is thus a misleading oversimplification to assert that the sigmoidal rate behavior is only observed at low glucose concentrations (below half-saturation), as is sometimes claimed, for example in a recent review.[22] Of course, if only a few observations are made, and these are mainly at high glucose concentrations, the cooperativity may pass unnoticed. For example, nobody would think that the first nine points (at high glucose concentrations) of the Eadie-Hofstee plot in Figure 3.2a fitted a straight line; it is obviously a curve and a plot of residual error[30] made after fitting a straight line would certainly confirm this. However, if instead of nine points there were only three or four in the same range one could easily be misled into thinking a straight line gave a reasonable fit, especially if there was a substantial amount of experimental error in the data. I must emphasize that insisting that the cooperativity is not "just" present at low glucose concentrations is more than the semantic or pedantic point that it may seem, because the property has important consequences for the physiological role of hexokinase D, as will be discussed in Chapter 6.

As hexokinase D is cooperative and does not follow Michaelis-Menten kinetics, it is wrong and misleading to refer to a K_m for glucose. The correct kinetic equation is not

$$\mathrm{v} = \frac{V_{\mathrm{max}}[\mathrm{Glc}]}{K_{\mathrm{m}} + [\mathrm{Glc}]} \tag{1}$$

but approximates to

$$\mathrm{v} = \frac{V_{\mathrm{max}}[\mathrm{Glc}]^h}{K_{0.5}^h + [\mathrm{Glc}]^h} \tag{2}$$

in which $K_{0.5}$ is the half-saturation concentration (note that raising it to the same power as the glucose concentration in the equation ensures that it keeps the dimensions of a concentration and thus has an easily understood physical meaning), and h is the Hill coefficient. Although the Hill equation is not based on a physical model (and thus needs to be replaced by more complicated kinetic equations when the mechanism of action is to be analyzed) it gives a good description of the kinetic behavior of hexokinase D and many other cooperative enzymes, and its parameters are very useful for descriptive and comparative purposes. By contrast, meaningful parameters cannot be obtained by fitting Equation (1), not even V_{max}; even though it appears in both equations its value will be different, possibly very different, depending on which equation is fitted. Unfortunately potentially interesting and important data have often been presented in ways that make it impossible to appreciate their significance. For example, a recent review[22] tabulates V_{max}, "K_m" and h values obtained with various mutants of hexokinase D associated with diabetes mellitus, but mixes parameter values obtained in two different (and incompatible) ways in the same table, the V_{max} and K_m values being obtained with Equation (1) and the h values probably from Equation (2) with $K^h_{0.5}$ replaced by K_m (the equation actually used cannot have been the one stated, which could not have given the parameter values listed). Note that even apart from the inconsistency of using the Michaelis constant in an equation that is not the Michaelis-Menten equation, writing $K^h_{0.5}$ as K_m produces a parameter with dimensions dependent on the value of h, which cannot be a concentration unless h = 1.

Figure 3.3, calculated from recent data of the same group on wild-type hexokinase D,[31] illustrates these points. Note in particular that

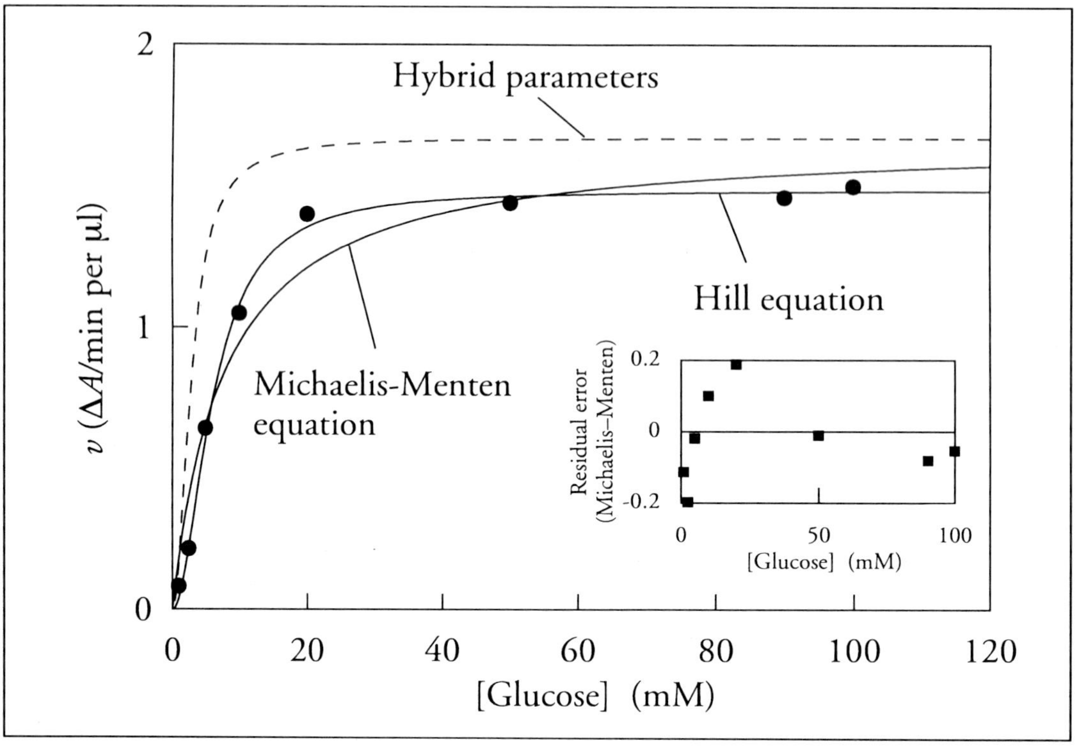

Fig. 3.3. Kinetic data for hexokinase D fitted in various ways. The data from Fig. 4 of Reference 31, in which the rate was measured as an increase in absorbance A, were fitted to the Michaelis-Menten equation and to the Hill equation (continuous curves). The broken line labelled "Hybrid parameters" illustrates the effect of inserting into the Hill equation the Michaelis-Menten parameters obtained by fitting the Michaelis-Menten equation together with the Hill coefficient obtained by fitting the Hill equation directly. Note that this gives a curve very different from the data points. The inset shows a plot of residual error for the curve obtained by fitting the Michaelis-Menten equation. Notice that the systematic deviations are much more obvious in the residual plot than in the original direct plot.

combining parameters obtained in two incompatible ways results in a hybrid curve that bears no resemblance to the experimental data and parameters obtained in this way do not provide any meaningful description of the kinetic properties of the enzyme. The curve obtained by fitting the Michaelis-Menten equation is less obviously objectionable, as it appears at first sight to fit the data reasonably well. However, the residual plot[30] given as an inset in Figure 3.3 shows obvious systematic deviations and, more important, the values of V_{max} and $K_{0.5}$ for the Hill equation differ substantially from the values of V_{max} and K_m estimated from the Michaelis-Menten equation. Of course, if the only purpose is to have a rough indication of the difference between mutants the Michaelis-Menten equation may provide this, but no quantitative conclusion would be obtainable, despite the fact that such quantitative conclusions are certainly needed if functions are to be assigned to specific amino acid residues.

Cooperative kinetics with glucose: a well preserved feature through evolution

As we saw in Chapter 1, hexokinase D exists in the livers of many mammals, reptiles and amphibians, though not all. Thus it was interesting to explore whether the sigmoidal rate behavior of rat and hamster hexokinase D with respect to glucose was also present in the enzyme from other species. As illustrated by Table 3.1, the enzyme from all the vertebrate species studied shows the same kinetic behavior as the rat enzyme,[11,18] with essentially the same Hill coefficient (1.4-1.7) throughout. Thus the sigmoidal saturation function appears to be a

Table 3.1. Kinetic characteristics of liver hexokinase D in different species[1]

Animal	h	$K_{0.5}$ (mM)
Mammalia		
Akodon *(Akodon olivaceus)*	1.4	3.0
Dog *(Canis familiaris)*	1.5	3.2
Pig *(Sus scrofa)*	1.5	3.8
Cururo *(Spalacopus cyanus)*	1.4	4.4
Coypu *(Myocastor coypus)*	1.4	6.2
Hamster *(Mesocricetus auratus)*	1.6	7.5
Rat *(Rattus norvegicus)*	1.6	7.5
Guinea pig *(Cavia porcellus)*	1.6	8.5
Reptilia		
Chilean turtle *(Geochelone chilensis)*	1.6	5.2
Amphibia		
Chilean frog *(Calyptocephalella caudiverbera)*	1.6	1.5
Clawed frog *(Xenopus laevis)*	1.7	2.6
Horned frog *(Ranita cornuda)*	1.7	2.6
Axolotl *(Ambistoma mexicanum)*	1.6	3.4

1 Data of Cárdenas and coworkers.[18]

common characteristic of vertebrate hepatic hexokinase D, and its preservation through evolution gives further support to the idea that it is of physiological importance.

It is noteworthy that amphibian hexokinases D maintain this sigmoidicity despite differing from the mammalian enzyme in immunological properties, to the extent that they do not cross react with the rat enzyme.[32,33] The turtle enzyme also requires a special comment because it was first described as showing Michaelis-Menten kinetics,[11] though later work showed without any doubt the same sigmoidal kinetics with respect to glucose and mannose as the rat enzyme. The discrepancy could result from the different seasons in which the experiments were done, as hexokinase D is not detected during the hibernation of the turtle, when the liver changes its appearance dramatically the first observations were, in fact, made with animals in a period close to hibernation.

The values of $K_{0.5}$ in different vertebrates vary rather more than the Hill coefficients, with a range of about 1-8 mM glucose at pH 7.5, with the amphibians consistently showing the lowest values. If there is any trend at all in the Hill coefficient it is also to show slightly higher values in amphibians.

The evolutionary preservation of the cooperative kinetics of vertebrate hexokinase D is paralleled by the presence of the hexokinase D regulatory protein in all the species where hexokinase D itself exists, as we shall see in Chapter 7.

KINETICS WITH RESPECT TO ATP

In enzyme reactions using two substrates one needs to ask whether the type of kinetics is similar for both of them; very often there are differences. For example, hexokinase C shows inhibition by excess of glucose but apparently Michaelis-Menten behavior with respect to MgATP.[34,35] Similarly, wheat hexokinase L_I shows negative cooperativity with respect to glucose but no deviations with respect to MgATP.[36]

The first studies of hexokinase D showed Michaelis-Menten behavior in relation to MgATP,[4-6] and unlike the case of glucose there was no suggestion of possible deviations. Nonetheless, these studies did raise some questions: (i) as mentioned previously, the enzyme had a low specific activity; (ii) the work had been done before the kinetic cooperativity with respect to glucose had been recognized; and (iii) as the true substrate is $MgATP^{2-}$, and as this ion does not exist in isolation but as a component of a mixture, the relative concentrations of Mg and ATP are crucial when studying MgATP, to ensure a constant proportion of MgATP at the different ATP concentrations. The earlier studies did not deal with this problem, which is discussed below in the context of the assay, and some of the current studies of hexokinase D do not do so either, which complicates interpretation of the results.

Storer and Cornish-Bowden made a detailed kinetic study with a pure enzyme of high specific activity.[12] They found that when the rate

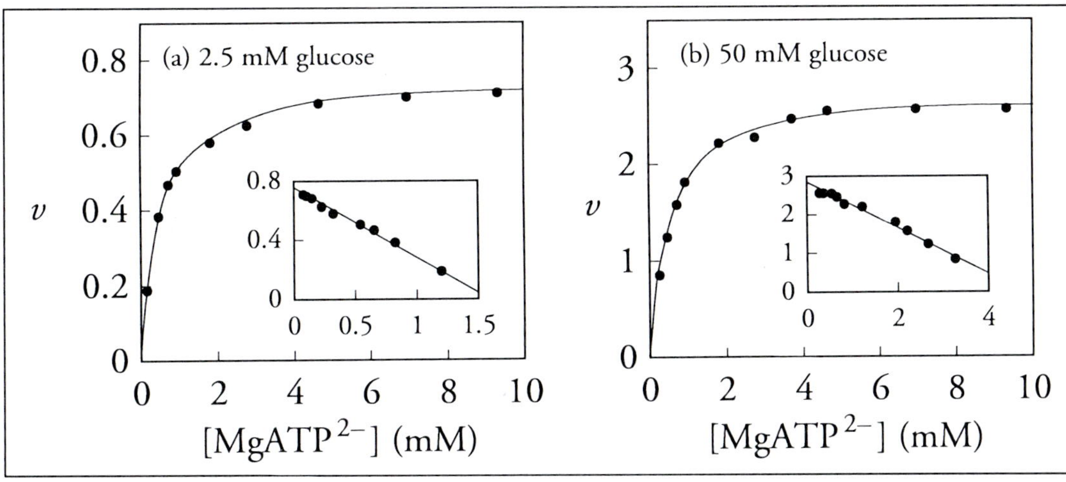

Fig. 3.4. Lack of cooperativity with respect to ATP. Data from Fig. 4 of Reference 12 have been recalculated to show the rate as a function of $MgATP^{2-}$ concentration, at (a) 2.5 mM glucose and (b) 50 mM glucose. Rates were measured at 30° C in glycylglycinate buffer at pH 8.0 in the presence of 1 mM excess $MgCl_2$ over the total ATP concentration. The original figure was presented as plots of $[MgATP^{2-}]/v$ against $[MgATP^{2-}]$ (Hanes-Woolf transformation), and included three additional plots at 10 mM, 5 mM and 1 mM glucose; none of the lines showed any perceptible deviation from linearity, as illustrated here by the straight lines in the corresponding Eadie-Hofstee plots shown as insets. The axes in these insets (unlabeled to avoid overcrowding) are as follows: abscissa, $v/[MgATP^{2-}]$; ordinate, v.

was measured as a function of the MgATP concentration at several fixed glucose concentrations, Michaelis-Menten kinetics were observed in all cases, validating the early studies (Fig. 3.4). At high glucose concentrations a K_m of 0.64 mM ATP (0.55 mM MgATP) was estimated, slightly smaller than the value of 1mM ATP (0.86mM MgATP) obtained by me at the same pH and at 100 mM glucose, as well as the values of 0.73 mM ATP and 1.0 mM ATP reported previously.[4,6] An important point to remember is that the cooperative kinetics with respect to glucose mean that one cannot accurately obtain the value of K_m for ATP by use of the ordinary secondary plots normally used in kinetic studies.

An interesting kinetic feature is that in all these studies the straight lines obtained in double-reciprocal plots at different glucose concentrations intersect in the third quadrant or on the abscissa axis. Even in this last case, the published graphs reveal a tendency to a lower apparent K_m at lower glucose concentrations. A completely different situation is observed when fructose or 2-deoxyglucose is used as substrate. In these cases the apparent K_m for ATP decreases as the concentration of fructose[37] or 2-deoxyglucose[38,39] increases, and in a double reciprocal plot the straight lines intersect in the second quadrant. In both cases the pattern obtained indicates a sequential mechanism in which both substrates must bind to the enzyme before the first product is liberated.

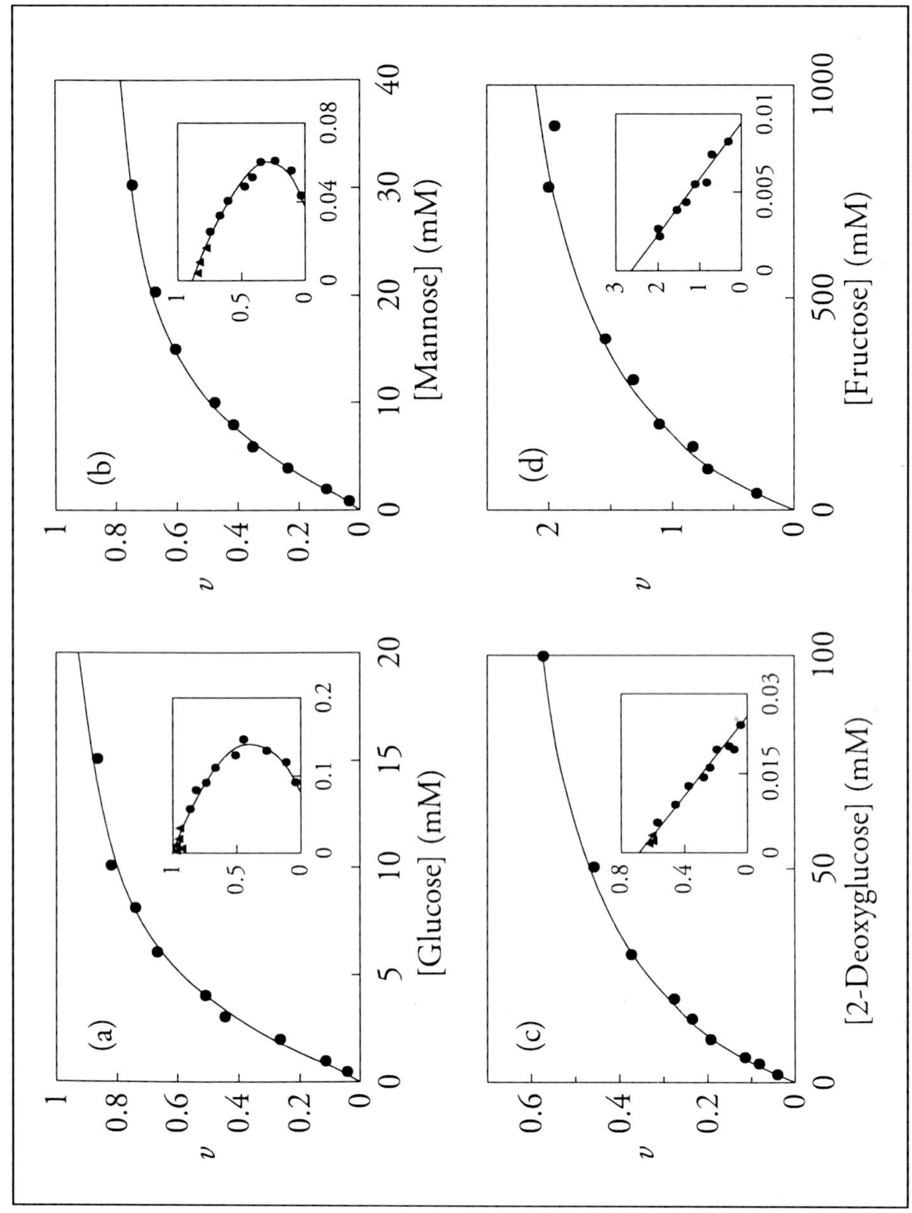

Fig. 3.5. Saturation functions of hexokinase D with various sugar substrates at pH 8.0: (a) glucose, (b) mannose, (c) 2-deoxyglucose, (d) fructose. Some of the points at the higher concentrations are omitted from the direct plots, to allow the use of scales in which the behavior at low concentrations is clearly visible. All points are shown in the corresponding Eadie-Hofstee plots shown as insets, the additional ones being shown as triangles (▲). The axes in these insets (unlabeled to avoid overcrowding) are as follows: abscissa, v/[sugar]; ordinate, v. Note that the curvature indicating cooperativity is essentially the same for glucose and mannose, but absent for 2-deoxyglucose and fructose. Data from Refs. 18, 37 and 40 were used for drawing the figure.

KINETIC BEHAVIOR WITH ALTERNATIVE SUGAR SUBSTRATES

We thought it important for understanding of the mechanism to determine if kinetic cooperativity also occurred with other sugar substrates, and it turned out that hexokinase D shows striking differences in kinetic behavior with different sugars. With mannose the sigmoidal dependence of rate with respect to the sugar concentration was similar to that with glucose, with a Hill coefficient of 1.5 (Fig. 3.5), and likewise the cooperativity with respect to mannose does not vary with pH, the degree of purification or the assay method.[18,40] In contrast, no cooperativity was observed with 2-deoxyglucose or with fructose (Fig. 3.5c,d),[18,37] and the Michaelis-Menten kinetics with 2-deoxyglucose suggested that it would be a useful substrate for studying the kinetic mechanism of hexokinase D.[38,39] The kinetic parameters with the four different sugar substrates at pH 8.0 were given in Table 2.2. The lowest $K_{0.5}$ is obtained with glucose (5 mM at pH 8.0) and the highest with fructose. As with glucose, the $K_{0.5}$ values for the other sugar substrates are lower than those reported previously,[4,5,9] probably for the same reasons.

In general, although many of the conclusions of the early studies[4,5,9] are still valid, at least qualitatively, the many uncertainties as to detail, especially cooperativity, make it highly misleading to continue quoting the parameters given in them while simultaneously referring to the Hill coefficient of 1.5; unfortunately this is done even in very recent reviews, such as that of Randle.[41] I must also emphasize that cooperativity is found with mannose as well as glucose, as this point also appears to have been misunderstood; for example, another recent review wrongly states that the reaction is noncooperative with respect to mannose.[22] Knowing that the cooperativity also exists with mannose is important for understanding current studies with mutants that are intended to establish structure-function relationships. For example, it appears that only mutations of residues predicted to form hydrogen bonds with glucose, such as Asn-204, Glu-290 and Glu-256, affect the cooperativity of the enzyme, as single mutations of any of these residues abolish the cooperativity,[22] and it will be interesting and useful to know how the same mutations affect mannose cooperativity.

ANOMERIC SPECIFICITY

The main difference between mannose and glucose as substrates for hexokinase D is the higher $K_{0.5}$ with mannose, the limiting rates and Hill coefficients being almost the same. This applies not only to the liver enzyme but also to that from pancreas.[42] As these two sugars differ in the anomer that predominates in aqueous solution (64% β-D-glucose, but 69% α-D-mannose),[43] one may enquire whether the difference in $K_{0.5}$ can be explained in terms of anomeric specificity. However, studies with pure anomers with both liver and pancreatic enzymes showed somewhat higher limiting rates for the β-anomers of both substrates, and much higher affinities for the α-anomers of both substrates.[42,44,45] Thus although anomeric specificity may be relevant to the small difference in limiting rate between glucose and mannose, it works in the wrong direction to explain the much larger difference in $K_{0.5}$.

FRUCTOSE KINETICS

The case of fructose deserves some special comments, as the low affinity of the enzyme for this substrate made it very difficult to estimate the saturation function, and the work of González and coworkers[4] gave the impression that the fructose activity might be due to a contaminating enzyme. Later on I clearly established that a pure preparation of hexokinase D does phosphorylate fructose, with Michaelis-Menten kinetic behavior (Fig. 3.5d) that is not affected by the degree of purification.[37] The apparent values of $K_{0.5}$ varied in the range of 276-380 mM fructose with different preparations of enzyme, values considerably smaller than those reported previously, greater than 0.8 M,[9] or even 2 M.[5] As with glucose, $K_{0.5}$ decreases on passing from pH 7.5 to 8.0 with no effect on the Hill coefficient of 1.0. Something very striking with fructose, however, was that despite the high value of $K_{0.5}$ the limiting rate V_{max} was more than double that obtained with glucose (Table 2.2), in support of an early observation of Parry and Walker.[5]

Initial rate studies with fructose show results compatible with a sequential mechanism with ordered addition of substrates or at random in rapid equilibrium, and in the following equation the values of the kinetic constants are $K_{m,Fru}$ = 174 mM, $K_{i,Fru}$ = 1.03 M, $K_{m,MgATP}$ = 0.35 mM:

$$v = \frac{V_{max}[\mathrm{Fru}][\mathrm{MgATP}]}{K_{i,FRU}K_{m,MgATP} + K_{m,MgATP}[\mathrm{Fru}] + K_{m,Fru}[\mathrm{MgATP}] + [\mathrm{Fru}][\mathrm{MgATP}]}$$

(The ratio $K_{i,Fru}K_{m,MgATP}/K_{m,Fru} = 2.06\mathrm{mM}$ is equivalent to $K_{i,MgATP}$ if a rapid-equilibrium random mechanism applies.) As with 2-deoxyglucose, the secondary plots are linear, in agreement with the hyperbolic characteristics of the kinetic curve.

DEPENDENCE OF THE COOPERATIVE KINETICS ON THE CONCENTRATION AND IDENTITY OF THE NUCLEOTIDE SUBSTRATE

(A) No cooperativity at low MgATP concentration

As hexokinase D is a monomeric enzyme even under assay conditions (see Chapter 6), the explanation of the cooperative behavior cannot reside in interactions between subunits. It could, however, be caused by the mechanism of the reaction. Ferdinand had pointed out that in a reaction with two substrates, such as that of hexokinase D, cooperative behavior could arise from the relative values of the kinetic constants,[46] as we shall see in Chapter 6. In such a mechanism the degree of cooperativity with respect to the variable substrate would depend on the concentration of the fixed one. Therefore it was important to investigate whether or not the degree of glucose cooperativity depended on the concentration of MgATP. When the experiments were done it was found possible to decrease the cooperativity by decreasing the MgATP concentration,[12] and that sufficiently low concentrations of MgATP would abolish it,[18] as shown in Figure 3.6. Together with the decrease in Hill coefficient a decrease in $K_{0.5}$ was also observed.[18] At high concentrations of MgATP, on the other hand, $K_{0.5}$ and h tend to remain constant. The cooperativity of Chilean frog hexokinase D also appears to depend on the MgATP concentration as the Hill coefficient of 2.3 decreases to 1.8 when the MgATP concentration is lowered from 10 to 1 mM. In this case there appears to be no effect on $K_{0.5}$.[47]

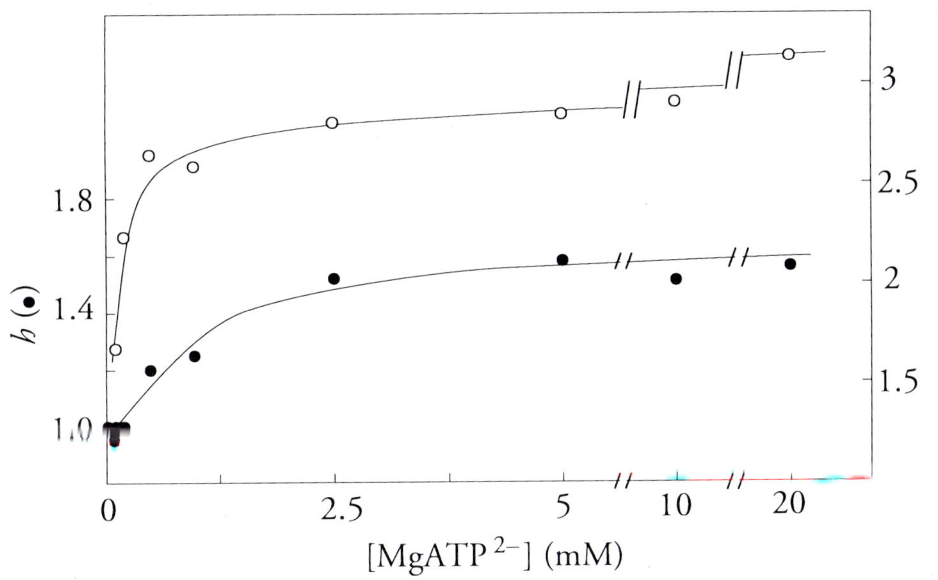

Fig. 3.6. Effect of MgATP on the kinetics of hexokinase D with respect to glucose. The Hill coefficient and the half-saturation concentration both decrease as the MgATP concentration decreases, with the Hill coefficient possibly approaching a value of 1 (no cooperativity) in the absence of MgATP. Redrawn from reference 18.

With 2-deoxyglucose the saturation function instead remains hyperbolic over a wide range of MgATP concentrations, and $K_{0.5}$ increases as the concentration of the nucleotide substrate decreases.[38,39] The same is observed with fructose.[37] The results obtained agree in general with Ferdinand's proposals,[46] but there was no inhibition by MgATP with any of the sugar substrates, as predicted by his model. This is further discussed in Chapter 6.

These results have a physiological interest beyond their mechanistic importance, because they mean that in the cell the concentration of ATP (estimated to be about 1 mM) is high enough to allow cooperativity in glucose phosphorylation, in agreement with the sigmoidicity of glucose utilization in vivo.[24,26] My experiments and those of Cornish-Bowden's group were done at pH 8.0. At this pH, 1 mM MgATP is a concentration where the sigmoidicity has started to decrease (Fig. 3.6), but at pH closer to 7.0 it appears that the sigmoidicity is still high at 1 mM MgATP. Thus Van Schaftingen, working with purified rat liver hexokinase D at pH 7.1, as we shall see in Chapter 7, obtained Hill coefficients of about 1.5-1.6 at 1 mM MgATP (e.g. reference 48). However, the interpretation of their results could be more complex, as will be discussed shortly.

(b) No cooperativity with MgITP as phosphate donor

As mentioned in Chapter 2, ITP appeared from the early studies to be a poor analog of ATP as phosphate donor in hexokinase reactions, and hexokinase D was no exception.[9,49] This was confirmed later when a pure enzyme of high specific activity was used to make a detailed kinetic study of hexokinase D with ITP as substrate, with the idea of getting insight into the molecular mechanism of the cooperativity, as described in Chapter 6.[50] A very interesting result that came from these studies is that when ATP is replaced by ITP the positive cooperativity disappears, and $K_{0.5}$ for glucose increases as the MgITP concentration increases.[50] In contrast to MgATP, for which the increase in K_m as the glucose concentration increases is only obvious at low glucose concentrations, K_m for MgITP is strongly dependent on the glucose concentration over the whole range, increasing from 4.5 mM at 1 mM glucose to 11 mM at 50 mM glucose; this last value is about 24-fold higher than the value for MgATP at the same glucose concentration.

CONSTANT DEGREE OF COOPERATIVITY IN DIFFERENT CONDITIONS

Whatever might be the mechanism of cooperativity, there is a theoretical possibility of modifying it by changes in the assay medium, such as pH or ionic strength changes, or by subjecting the enzyme to treatments such as photooxidation or the presence of denaturing agents. In principle, any of these treatments could produce differential changes in the different conformations or the different binding sites, or it could perturb the interactions between subunits. As we shall see below, cooperativity proved to be a very constant feature that was not modified by treatments of this kind. In all the experiments mentioned below precautions were taken to ensure that any effects seen were not due to effects on the auxiliary enzyme or enzymes needed for the coupled assay.

Effect of pH and ionic strength

Hexokinase D has its maximal activity at about pH 8.[4,5,9,18,51] The degree of sigmoidicity is not modified by variations in pH in the range

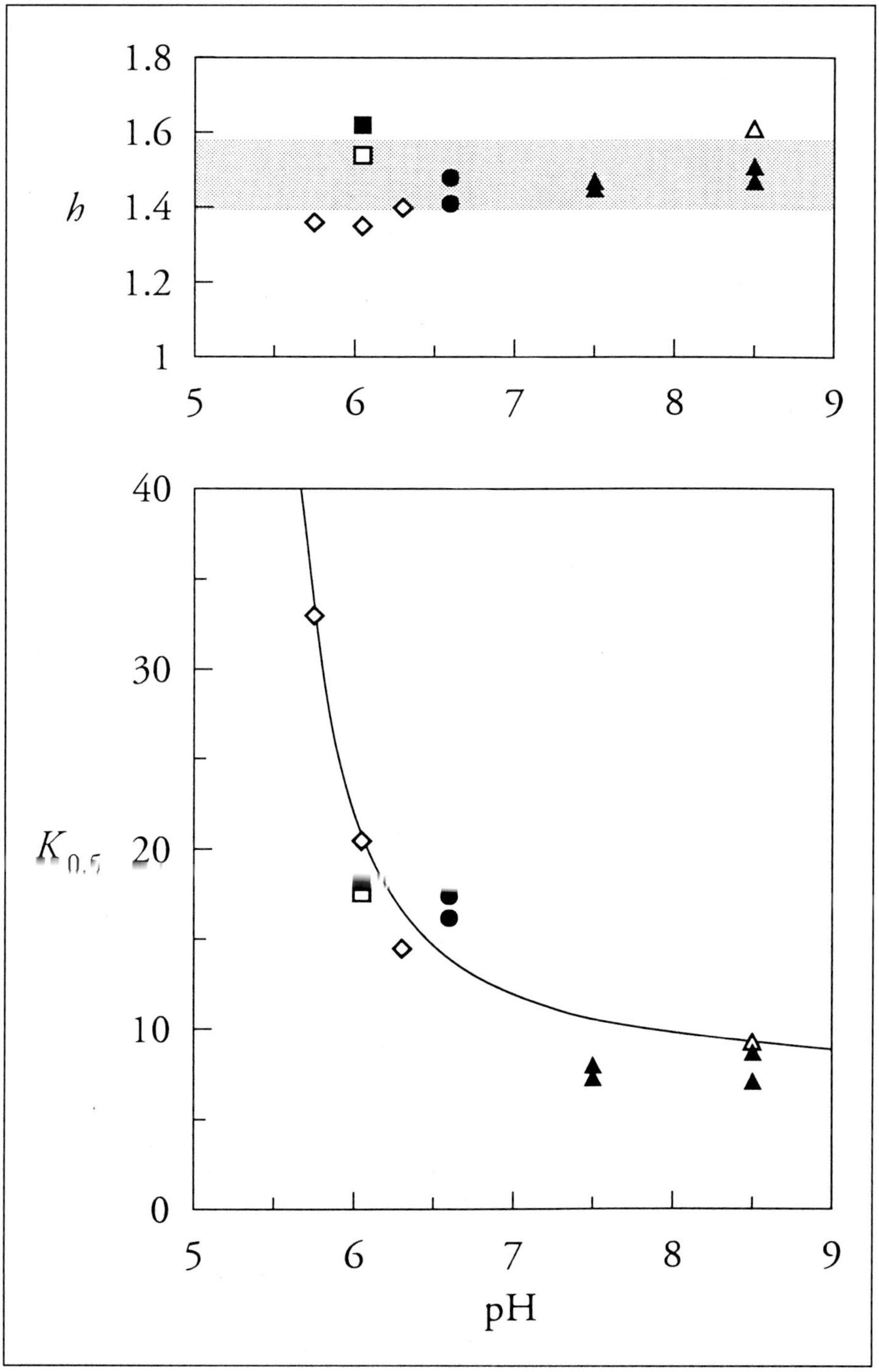

Fig. 3.7. Effect of pH on the kinetics of hexokinase D with respect to glucose. The Hill coefficient is essentially independent of pH whereas the half-saturation concentration decreases as the pH increases. The different symbols represent different buffers used. Reproduced from the author's doctoral thesis.[58]

5.7 to 8.5; in different experiments using different types of buffers the Hill coefficient was essentially constant at 1.48 ± 0.08.[18] On the other hand, in the same experiments there were large changes in the half-saturation concentration $K_{0.5}$, which increased greatly at pH values lower than 7.0, as can be seen in Figure 3.7. This variation of $K_{0.5}$ as a function of pH suggests that an acid group with a pK of about 7, perhaps the imidazole group of a histidine residue, could be involved in the binding of glucose. In agreement with this idea, a plot of p$K_{0.5}$ against pH according to Dixon[52] reveals a group of pK about 7 involved in the enzyme-glucose complex.

This constancy of the sigmoidicity of hexokinase D in relation to pH contrasts with the behavior of other cooperative enzymes, such as phosphofructokinase and pyruvate kinase, both of which are markedly affected by pH.[53,54] Cattle pancreatic ribonuclease, which is monomeric like hexokinase D and shows sigmoidicity with some synthetic substrates, resembles these more "classic" examples of cooperative enzymes in that its sigmoidicity can be abolished by decreasing the pH below 6.6.[55] This enzyme has allosteric sites for binding synthetic substrates that induce a conformational transition,[56] and a pH-dependent partition between at least two conformational states differing in activity was postulated.[57] There are no corresponding pH effects on the cooperativity of hexokinase D, so in this case the conformational distribution cannot be pH-dependent.

Effects of the concentrations of K^+ and Na^+

Increasing the concentration of K^+ or Na^+ up to 0.4 M appears to have no significant effect on the degree of cooperativity of hexokinase D with respect to glucose, even though V_{max} and $K_{0.5}$ decrease slightly, by about 20-30%. Because of the effect on $K_{0.5}$, both cations can act in opposite directions at different glucose concentrations, as activators at low glucose concentrations and as inhibitors at high concentration.

These kinetic studies suggested that K^+ and Na^+ were equivalent in their effects on the kinetic parameters. However, they are not equally effective for stabilizing the enzyme, as we shall see shortly, nor are they equivalent in studies in hepatocytes, as cells incubated with K^+ have a significantly smaller Hill coefficient and $K_{0.5}$ for glucose than cells incubated with Na^+.[24] These results will be discussed in Chapter 7.

It has been known for a long time that K^+ is a stabilizing agent for hexokinase D,[4,5,7-9,20] and for this reason it is used during the purification of the enzyme. Na^+ cannot replace K^+ as a stabilizing agent, at least in the first steps of purification, and the yield with Na^+ is about 50-60% of the yield with K^+.[58] Furthermore, the thermal stability of the enzyme varies with the cation added; hexokinase D prepared in the presence of Na^+ is more susceptible to inactivation at 37° than enzyme obtained in the presence of K^+ (Fig. 3.8), and this inactiva-

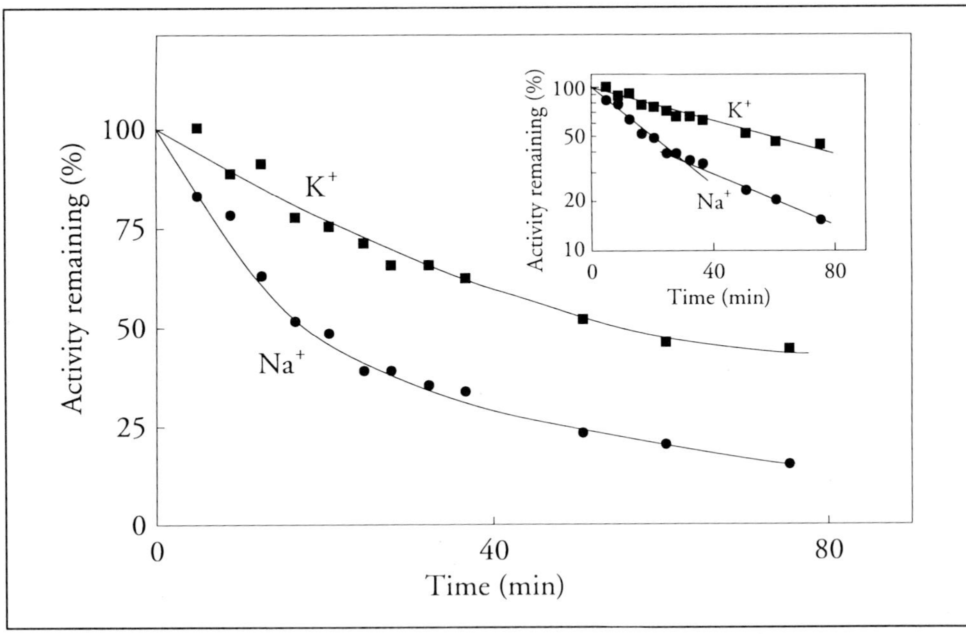

Fig. 3.8. Effects of metal ions on thermal stability. Hexokinase D loses activity more rapidly at 37° in the presence of Na^+ (●) than in the presence of K^+ (■), and the inactivation in the presence of Na^+ does not follow first-order kinetics. See the text for further details. Reproduced from the author's doctoral thesis.[58]

tion does not follow first-order kinetics, as appears to be the case with the enzyme prepared with K^+. All of this confirms the difference between Na^+ and K^+ as stabilizing agents.

Effect of thiol protectors and modifiers

Mammalian and yeast hexokinases are sensitive to the modification of thiol groups, which produces inactivation;[59] this was also observed also with hexokinase D in the very early studies with organomercurial compounds and alkylating agents.[6,7,9]

Reagents that reduce disulfide bridges, such as dithiothreitol and 2-mercaptoethanol, act to stabilize hexokinases. The presence of dithiothreitol as a stabilizing agent is very important in the assays of hexokinase D in vitro, as it prevents the inactivation that is especially prevalent at low glucose concentration. Dithiothreitol not only prevents inactivation, but can also reverse it, totally or partially. Even an enzyme that has completely lost its activity over a period of time can be partially reactivated (albeit to a small extent) by preincubation with dithiothreitol. In all these cases the degree of cooperativity is not altered, in experiments done at pH 8.0 and 7.5.[18]

An interesting observation was made by Niemeyer's group, who used the protection of hexokinase D afforded by substrates and products against inactivation by 5,5'-dithiobis(2-nitrobenzoic acid) (DTNB) as a probe of the interaction of those ligands with the enzyme.[60] They observed that after storage for several weeks, hexokinase D lost its capacity to be inactivated by DTNB and only after that this had happened did the enzyme begin to lose activity progressively (Fig. 3.9).[61] The catalytic function of hexokinase D and its susceptibility to DTNB inactivation were partially recovered by incubation of the enzyme with dithiothreitol. The degree of recovery decreased progressively with time. Furthermore, in enzyme partially inactive and resistant to DTNB, which must have some thiol groups oxidized, the values of the Hill coefficient and half-saturation concentration were similar to those for the native enzyme.[61] This result contrasts with what is observed with frog hexokinase C, which loses its inhibition by excess glucose after modification of thiol groups.[34]

The separation between catalytic activity and susceptibility of hexokinase D to DTNB inactivation indicates that the fast-reacting thiol groups of hexokinase D do not participate directly in the catalytic mechanism, but that they are important for stabilizing a conformation which can exist in their absence, but becomes unstable. I have always found these results very interesting and to the best of my knowledge such separation of effects has not been reported for other enzymes.

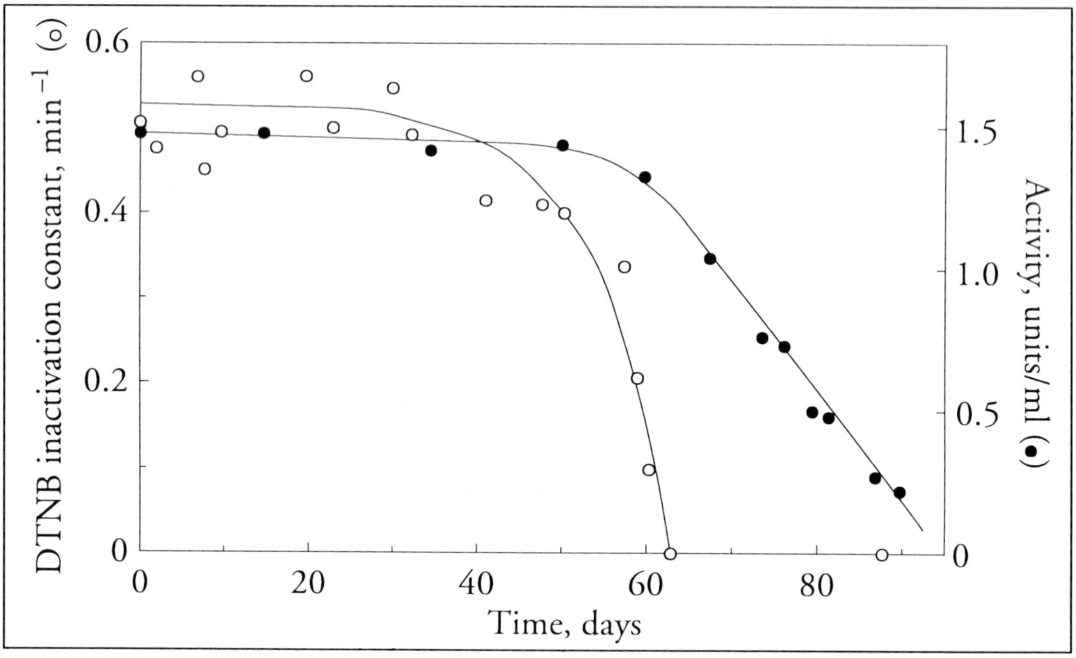

Fig. 3.9. Dissociation between catalytic activity and susceptibility to DTNB. Hexokinase D is initially sensitive to inactivation by DTNB, but the sensitivity decreases steeply after storage for a few weeks (○), after which time the enzyme begins a slow DTNB-independent loss of activity (●). Redrawn from Arch Biol Med Exper 1981; 14:177-182.

Something to bear in mind when one is working with a hexokinase D that is not 100% pure is that *N*-acetylglucosamine kinase is not susceptible to spontaneous oxidation and it is not activated by dithiothreitol. This enzyme is easily confused with hexokinase D, as discussed in Chapter 4, and it is a common contaminant of hexokinase D preparations. Consequently a very minor contamination by *N*-acetylglucosamine kinase may become progressively more important as hexokinase D loses activity with time and the activity of *N*-acetylglucosamine kinase remains more or less stable; it can thus happen that the final residual activity is not due to hexokinase D at all.[58]

Enzyme concentration

There are several examples of enzymes such as glutamate dehydrogenase that can self-associate to give oligomers with different properties from those of the monomers, as reviewed by Frieden.[62] As the fraction of oligomer then varies with the protein concentration the kinetic behavior ought to vary likewise, but this does not happen with hexokinase D; the rate varies linearly with protein concentration, with no change in kinetic parameters. This observation agrees with the results of gel filtration on Sephadex G-200, which show a constant distribution coefficient that corresponds to a monomer at every enzyme concentration (see Fig. 6.2).

Photooxidation with methylene blue

Photooxidation of hexokinase D with methylene blue at pH 7.0 inactivates it in a first-order process. However, the photochemical modification does not suppress or modify the cooperativity,[18] but only produces an increase in the half saturation concentration as the degree of inactivation varies between 15 to 75%. In contrast, frog hexokinase C (but not the rat enzyme) loses the inhibition by excess glucose after photooxidation,[34,35] a difference that agrees with the results for DTNB noted above.

This greater sensitivity of the frog hexokinase C to photooxidation than the corresponding rat enzyme led us to also study hexokinase D from a frog (*Calyptocephalella caudiverbera*). However, the experiments with this enzyme were not successful because it was too sensitive to the inactivation. Even methylene blue by itself in the dark for 10 min (a treatment that has no effect on rat hexokinase D), was sufficient to inactivate the frog enzyme about 50%. It appears that frog hexokinase D is much more sensitive to photooxidation than hexokinase C of the same species, which is only slightly inactivated under the same conditions.

Methylene blue modifies histidine residues in several enzymes,[63] although in some experimental conditions it can also modify other residues, such as cysteine, methionine, tryptophan and tyrosine. However, photooxidation with Rose Bengal, a dye with great selectivity for histidine, also had no effect on the cooperativity (Hill coefficient of 1.52) under conditions in which V_{max} decreased by 50%. The fact that photooxidation not only decreases V_{max} but also increases $K_{0.5}$ means that this process both decreases the number of catalytically active molecules and also generates altered enzyme molecules with a lower affinity for glucose but with cooperativity unchanged.

The photooxidation experiments suggest the participation of a histidine residue in the active site, which would be compatible with the variations observed in $K_{0.5}$ and V_{max} as a function of pH, as they suggest the presence of an ionizable group with a pK about 7.0.

Heating

Variations in temperature alter hydrophobic interactions in proteins, which could produce conformational changes and alter kinetic behavior. In the case of hexokinase D, heating for variable periods with different degrees of inactivation and subsequent studies at 30° only increased $K_{0.5}$ without altering the Hill coefficient.

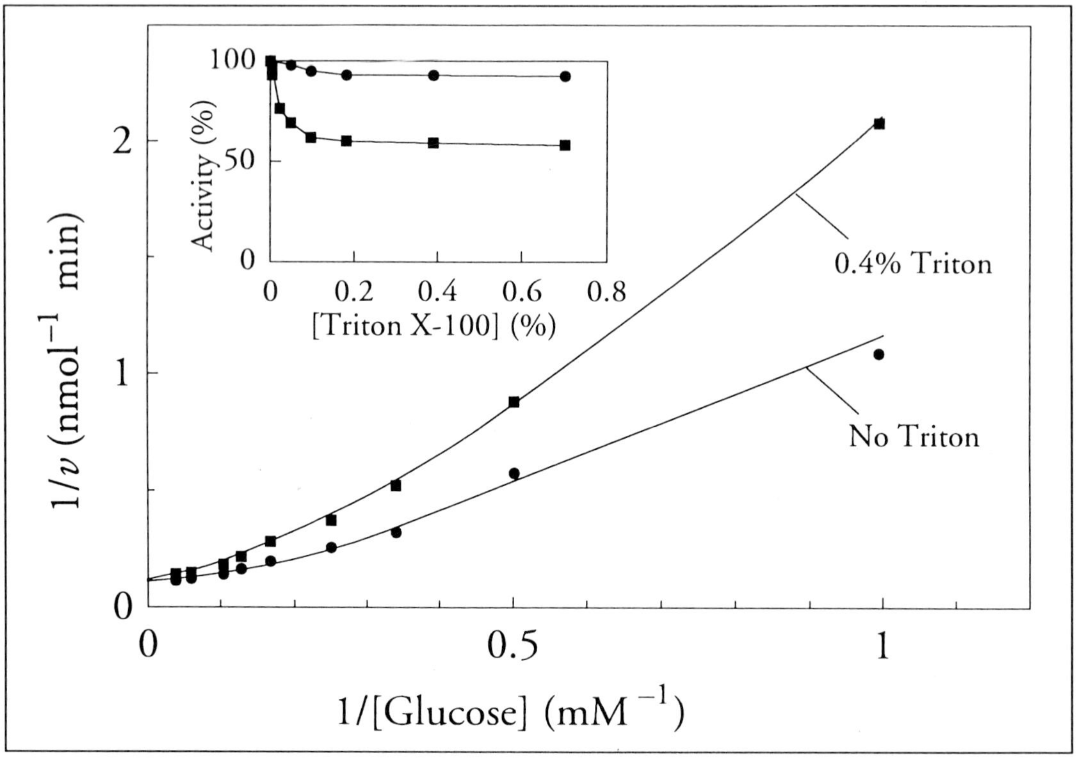

Fig. 3.10. Effect of a non-ionic detergent on the kinetics. The cooperativity and limiting rate of hexokinase D are unaffected by Triton X-100 up to 0.5% (five times the critical micelle concentration), but the half-saturation concentration was increased. As shown in the inset, the inhibition by Triton X-100 was incomplete, i.e. the activity did not approach zero at high detergent concentrations. Previously unpublished data of the author.

Effect of Triton X-100

Triton X-100 is a nonionic detergent that forms micelles at concentrations over 0.1%. However, on both sides of this concentration, and as high as 0.5%, its presence in the assay medium affects neither V_{max} nor the degree of cooperativity of hexokinase D, though the value of $K_{0.5}$ increases (Fig. 3.10). As with acyl-CoA (Fig. 7.1), the inhibition is incomplete, as may be seen in the inset to Figure 3.10, where a glucose concentration close to $K_{0.5}$ was used. No attempts were made in this case to see whether changing the critical micelle concentration might alter the degree of inhibition, as in the studies of long-chain acyl-CoAs that will be discussed in Chapter 7. Despite this lack of effect on V_{max} the enzyme becomes inactivated during the assay at high concentrations (over 0.1%) especially at glucose concentrations lower than 8 mM,[58] probably because of the presence of micelles. Consequently precautions need to be taken to avoid artifacts when the rate is estimated.

Effect of urea

The presence of urea in the assay medium produces an instantaneous inhibition of hexokinase D activity that increases with the urea concentration.[7,18,64] At concentrations higher than 3 M the activity is zero. In contrast to the experiments with Triton X-100, and in spite of the large inactivation effect, the enzyme is very stable during the assay at different urea concentrations, for times as long as 20 min, even at low glucose concentrations. The effects of urea at concentrations of 1 M, 3 M or 8 M are reversible in rat and dog hexokinase D, and the reversibility appears to be instantaneous, i.e. at these concentrations the final activity depends on the urea concentration in the assay, not that during the preincubation. In contrast, at intermediate high concentrations of urea of about 5 M the effect appears not to be reversible.[58,64]

Urea decreases V_{max} and increases $K_{0.5}$, but it appears not to modify the degree of cooperativity.[18] These results contrast with what has been reported for wheat-germ hexokinase L_I which is also monomeric and shows deviations from Michaelis-Menten behavior (negative cooperativity). A small urea concentration (50 mM), which has almost no effect on rat or dog hexokinase D,[18,64] is sufficient to desensitize the wheat-germ enzyme, eliminating the negative cooperativity with respect to glucose and decreasing the activity by more than 50%.[36]

In summary, the presence of Triton X-100 or urea in the assay mixture, as well as changes in pH or the concentration of Na^+ or K^+, do not modify the degree of cooperativity. Equally, chemical modification of some residues (photooxidation of histidine, oxidation of cysteine), which like the other conditions mentioned above could disturb the conformational distribution, did not affect cooperativity; neither did heating. In all these cases, which include treatments before the assay, during it, or both, the results were in general similar; there were major alterations in the enzymatic activity, a decrease in V_{max} (in the case of Triton X-100 only, there was no variation), and a change in the half-saturation concentration (decreased by K^+ and Na^+, in other cases increased).

The alterations in $K_{0.5}$ indicate that in all these conditions, in addition to a possible complete inactivation of some enzyme molecules, which could explain the decrease in V_{max}, there were also conformational alterations in molecules that retained partial or complete enzymatic activity. These conformational modifications appeared, however, not to interfere with the specific effect of glucose, which as we shall see in Chapter 6 consists of inducing a conformational change responsible for the cooperative kinetics. The susceptibility of the parameters $K_{0.5}$ and V_{max} to respond to variations in the assay medium or to different kinds of pretreatment contrasts strikingly with the constancy of the Hill coefficient in the same conditions. Apparently hexokinase D retains its non-Michaelian behavior as long as retains activity. In other words, the cooperativity appears to be tightly linked to the catalytic cycle.

Does all this mean that in no condition is the cooperativity decreased (or increased)? The answer is no. In Chapter 6 we shall see that the degree of cooperativity with respect to glucose not only depends, as already discussed, on the concentration of the second substrate, MgATP, but it can also be abolished by the presence of competitive inhibitors. Likewise, the presence of glycerol in the assay mixture, or the replacement of H_2O by ^2H_2O, produces a decrease in the degree of cooperativity. These effects are more easily understood in relation to the models that have been proposed for the cooperativity of hexokinase D, and they will accordingly be described in Chapter 6.

ASSAY AND PURIFICATION

Assay conditions needed for a stable enzyme

One reason why it took us a long time to publish the kinetic cooperativity of hexokinase D with respect to glucose was the concern that it could be an artifact, as hexokinase D is an unstable enzyme that could get partially inactivated during the assay with the degree of inactivation dependent on the glucose concentration. As a matter of fact, the enzyme obtained at one of the last stages of purification used during that period was unstable and at low glucose concentration became inactivated during the assay. If the degree of inactivation increased as the glucose concentration decreased this would produce an artifactual sigmoidal curve that was not an intrinsic property of the enzyme.

For this reason much effort was devoted to find conditions to avoid inactivation, such as the addition of dithiothreitol and K^+. Dithiothreitol added at concentrations of 2.5 mM could prevent the inactivation, and dithiothreitol at 10 mM could reverse inactivation that had already happened. Equally, at pH 7.5, concentrations above 30 mM of K^+ help to prevent inactivation at low glucose concentration. As this is an intracellular cation with a steady-state concentration of about 160 mM in the hepatocyte,[65] a concentration of 100 mM did not appear excessive. It is true that this implies a Cl^- concentration higher than the physiological value of about 30-50 mM.[66] However, as noted above variations in the concentrations of K^+ and Na^+ do not affect the degree of cooperativity if one pays attention to the problem of inactivation.

To make sure that there was no inactivation during the assay, progress curves were recorded for periods long enough that any inactivation which occurred would be detected. The need for this precaution has increased with the appearance of advanced modern spectrophotometers that provide rate values directly, because if there is any problem of inactivation the progress curve will deviate from linearity over the whole time course, and one cannot obtain an adequate estimate of the initial rate by using the built-in software to fit a straight line to part of the data. Unfortunately not all manufacturers appear to be aware of this and supply user manuals that recommend straight-line regression for fitting points that deviate in an obvious way from a straight line.[67]

In the light of these worries about the possibly artifactual nature of the cooperativity, the report that glucose phosphorylation was also sigmoidal in isolated hepatocytes[24] was very reassuring, because it indicated that the enzyme had properties in physiological conditions similar to those of the purified enzyme, suggesting that these properties were meaningful in the cell.

Vandercammen and Van Schaftingen have argued that the use of pH 8.0 and 100 mM KCl in the assay medium could lead to misleading results, as the physiological pH is pH 7.2 and the Cl^- concentration is only 25 mM.[48] However, it depends on what one is looking for. If the idea is to understand the kinetic mechanism of the purified enzyme, these conditions are quite valid because it was established that variations in pH or K^+ concentration did not affect the degree of cooperativity. Furthermore although the Cl^- concentration in hepatocytes is about 30-50 mM,[66] the K^+ concentration is much higher, about 160 mM.[65] From that point of view, therefore, use of 25 mM KCl is not very physiological either, and this concentration of K^+ is rather low for maintaining hexokinase D in a stable state. In addition, in studies of the pure enzyme loss of activity during the assay can generate artifacts such as apparent increases in the degree of cooperativity. Finally, at pH 7.2 formation of the ion $MgATP^{2-}$ is less favored than at pH 8.0, so that its concentration will be significantly less than the total ATP concentration. This is especially a problem if equimolar Mg^{2+} and ATP are used, because of nonproportionality between the concentrations of $MgATP^{2-}$ and total ATP. If this is not taken into account it is easy to generate a spurious appearance of deviations from Michaelis-Menten kinetics due to the dependence of rate on $MgATP^{2-}$ concentration. On the other hand, if the objective is to study interactions between hexokinase D and other types of protein or metabolites that may interact with it in vivo, one can only agree that working at pH 7.2 is important, though precautions are still needed to avoid the problems mentioned.

The problem of instability during the assay is especially important for kinetic studies of artificial mutants, and needs to be borne in mind during the present period in which mutants are being produced and kinetically characterized to obtain a better understanding of metabolic illnesses like NIDDM (see for example references 22 and 68), as we shall see in Chapter 5. Such mutants are likely to be even less stable than the native enzyme under assay conditions, and hence more likely to generate artifacts, even if there are no obvious differences in stability in vivo.

Hexokinase assays

Spectrophotometric procedures

The standard methods for detecting hexokinase D activity are spectrophotometric procedures that measure the production of glucose 6-phosphate (Assay 1a) or ADP (Assay 2) with the use of auxiliary coupling enzymes to detect the change in absorbance of NADH or NADPH.[69] When other substrates such as fructose or mannose are used instead of glucose, the corresponding phosphate esters can be measured by Assays 1b and 1c, which use different coupling enzymes. Assay 2 can be used with any sugar substrate. As all of these methods involve the use of auxiliary enzymes, precautions should be taken to choose the correct amounts of them.[70] This is especially important with assays 1b and 1c because of the high K_m values of the isomerases used.

Assay 1a

The production of glucose 6-phosphate is detected by coupling it to the production of NAD(P)H by addition of glucose-6-phosphate dehydrogenase and excess of NAD(P):

$$\text{Glucose} + \text{MgATP} \rightarrow \text{glucose 6-phosphate} + \text{MgADP}$$

$$\text{Glucose 6-phosphate} + \text{NAD(P)} \rightarrow \text{6-phosphogluconate} + \text{NAD(P)H}$$

The NAD(P)H produced in the coupled reaction has an absorption peak at 340nm (molar extinction coefficient of 6220 M^{-1} cm^{-1}), which is absent in the initial oxidized compound.[71] The reduction can also be followed fluorimetrically, with a large increase in sensitivity. In liver extracts the presence of 6-phosphogluconate dehydrogenase introduces uncertainties as to the stoichiometry between the NADPH formed and the glucose phosphorylated, which can vary between 1 and 2 according to the degree of contamination, and can be a problem if yeast glucose 6-phosphate dehydrogenase is used, because it functions with NADP (as does the liver enzyme).

The assay can be improved by using a NAD-linked glucose 6-phosphate dehydrogenase as coupling enzyme, such as the commercially available enzyme from *Leuconostoc meserentoides*. NAD is not only much cheaper than NADP, but, more importantly, it is not a substrate for 6-phosphogluconate dehydrogenase. A standard procedure for measuring hexokinase D activity in extracts is to measure activities at 100 mM and 0.5 mM glucose, to detect the total hexokinase and the low-K_m hexokinases respectively; the difference between the total and low-K_m activity is then taken as a measure of hexokinase D. However, this is less straightforward than it may appear at first sight, and is discussed in Chapter 4.

Assay 1b

Production of fructose 6-phosphate can be followed by the reduction of NAD(P) in the presence of glucose 6-phosphate dehydrogenase and phosphoglucose isomerase:[71]

Fructose + MgATP → fructose 6-phosphate + MgADP
Fructose 6-phosphate → glucose 6-phosphate
Glucose 6-phosphate + NAD(P) → 6-phosphogluconate + NAD(P)H

Assay 1c

The production of mannose 6-phosphate can be determined in the presence of glucose 6-phosphate dehydrogenase, phosphoglucose isomerase and phosphomannose isomerase as coupling enzymes.[71]

Assay 2

The production of ADP is followed through the oxidation of NADH in the presence of pyruvate kinase and lactate dehydrogenase as coupling enzymes:[72]

Glucose + MgATP → glucose 6-phosphate + MgADP
MgADP + phosphoenolpyruvate → MgATP + pyruvate
Pyruvate + NADH → lactate + NAD

As this assay measures production of ADP it is important to bear in mind that even if ATP stock solutions are prepared daily they contain ADP as a contaminant. Consequently, if the reaction is started by addition of ATP, and no special care is taken when following the progress curve for a short time, the estimated initial rate will be larger than the correct value because of the contaminating ADP. The problem is much worse if the ATP solution is not fresh or is kept at an inappropriate pH. It can be eliminated by preincubating the ATP with the assay mixture.

Assay 2 is especially recommended for studying the dependence of the rate on the MgATP concentration, because the regeneration of ATP during the assay means that measurements can be made at a constant MgATP concentration even at the lowest values used, which is important for estimating the initial rate correctly. If assay 1 needs to be used it is still possible to maintain a constant MgATP concentration by using another ATP-regenerating system such as phosphocreatine and creatine kinase.

Specificity of spectrophotometric assays

Obviously Assay 1 is more specific than Assay 2 and is consequently better for crude extracts. Nonetheless, the glucose-6-phosphate dehydrogenases of yeast and liver are also capable of oxidizing 2-deoxyglucose 6-phosphate and fructose 6-phosphate, but with low activity.[58] This must be taken into account in experiments in which fructose or 2-deoxyglucose may also be present. Thus even if it may appear that glucose 6-phosphate dehydrogenase from different sources has a high substrate specificity this is not absolute and other phosphate esters may also be oxidized.[73,74,75] Furthermore, a human variant of glucose 6-phosphate dehydrogenase called Kremenchung exists that has a high affinity for 2-deoxyglucose 6-phosphate.[76] This lack of absolute specificity means that at low glucose concentrations an acceleration in the progress curve may be detected after a certain time if 2-deoxyglucose or fructose is present.

Radiometric assays

There are several types of radiometric assays using isotopically labelled glucose.[23,77,78] Some of these methods involve direct measurement of glucose 6-phosphate, either by binding it to Dowex columns followed by elution,[23] or by binding it to DEAE-cellulose filter disks,[77] so that it is possible to do the assay in the presence of glucose 6-phosphate to inhibit the low-K_m hexokinases. The radiometric assays have the further advantage of avoiding the use of coupling enzymes, but for kinetic experiments the spectrophotometric method remains the best.

Mg^{2+} concentration

The true nucleotide substrate of hexokinases is $MgATP^{2-}$ and, because it is not possible to have this as a single species since it is impossible to prepare a solution of pure $MgATP^{2-}$, the relative concentrations of Mg and ATP are very important for ensuring that more than 80% of the ATP is in the form of MgATP. Therefore it is important to state clearly how much Mg and ATP is being used and not to report only the claimed concentration of MgATP, as is often unfortunately done. Any solution that contains $MgATP^{2-}$ also contains numerous other ions: for example, an equimolar mixture of ATP and $MgCl_2$ at pH 7.0 contains appreciable proportions of $MgATP^{2-}$, ATP^{4-}, $HATP^{3-}$, Mg^{2+} and Cl^-, as well as traces of $MgHATP^-$, Mg_2ATP and $MgCl^+$.[79] The proportions of these species vary with the total ATP and $MgCl_2$ concentrations, the pH, the ionic strength and the concentrations of any other species (i.e. buffer components) that may be present.

The control and precise knowledge of the concentration of $MgATP^{2-}$ is especially important if one is studying the dependence of rate on

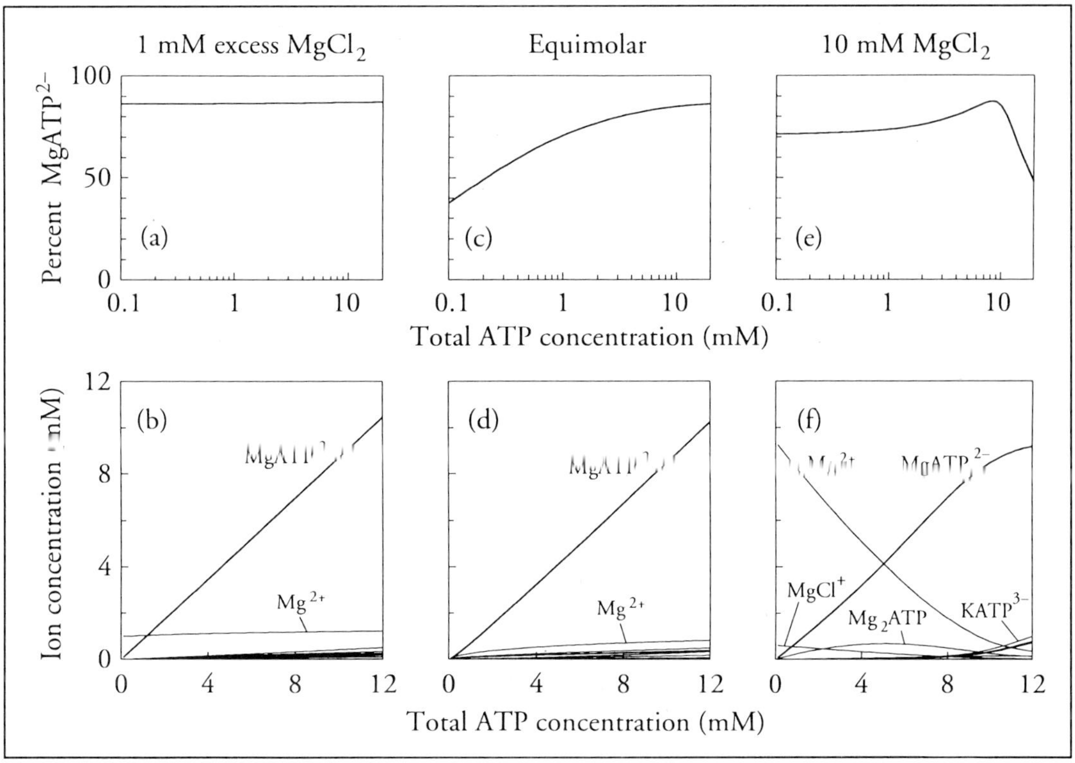

Fig. 3.11. Control of $MgATP^{2-}$ and free Mg^{2+} concentrations. If the total Mg^{2+} concentration is maintained in constant 1 mM excess over the total ATP concentration, a high and essentially constant proportion of the total ATP exists as $MgATP^{2-}$ (a), the concentration of free Mg^{2+} is small, constant, and similar to the value in cells, and other ions are present at very low levels (b). However, if total Mg^{2+} and total ATP are varied in equimolar concentrations the proportion of total ATP that exists as $MgATP^{2-}$ varies excessively (c), and the concentration of Mg^{2+} is far from constant (d). Maintaining total Mg^{2+} at a constant high value yields a satisfactorily high and constant proportion of total ATP existing as $MgATP^{2-}$, as long as the total ATP concentration is kept well below the total Mg^{2+} concentration (e), but the proportion falls steeply at higher total ATP concentrations, the concentration of free Mg^{2+} varies widely and concentrations of other ions may become important (f). The curves were calculated using the method of Storer and Cornish-Bowden.[79]

$MgATP^{2-}$, for example in different mutants of hexokinase D. A "good" design for obtaining a high and almost constant proportion of the ATP existing as $MgATP^{2-}$ and a nearly constant concentration of free Mg^{2+} is to keep the total $MgCl_2$ concentration in constant excess over the total ATP concentration. By contrast, the common practice of varying the total concentrations of ATP and $MgCl_2$ in constant ratio gives very unsatisfactory results, with large variations in the proportion of ATP that exists as $MgATP^{2-}$. Another design often used in the past was to maintain a constant total $MgCl_2$ concentration at all ATP concentration. This gives a reasonably constant and high proportion of total ATP present as $MgATP^{2-}$, as long as one is careful not to allow the ATP concentration to exceed the total $MgCl_2$ concentration, but it is accompanied by large variations in the free Mg^{2+} concentration that may cause difficulties in interpreting the results if the enzyme is, like hexokinase D, sensitive to free Mg^{2+}. These general features are illustrated in Figure 3.11.

Using an excess of about 5 mM $MgCl_2$ over total ATP gives a high constant proportion of $MgATP^{2-}$ (about 90%) over a wide range of ATP concentrations, but as hexokinase D is inhibited by Mg^{2+} it is better to decrease this excess to 1 mM; this results in only a small loss of efficiency, and in the range 0.1-10 mM of total ATP concentrations it gives an acceptably constant proportion (84-88%) of ATP present as MgATP. With Assay 1a this concentration of Mg^{2+} works well, but Assay 2 presents the special problem that pyruvate kinase, an essential component of the assay, requires at least 4 mM of free Mg^{2+} for activity;[80] in this case it is better to use an excess of 5 mM $MgCl_2$, even at the risk of inhibiting hexokinase D. The alternative of enormously increasing the quantity of auxiliary enzyme is unsatisfactory, not only because of the expense but, more importantly, because it leads to problems of contamination. For further details and discussion see references 79 and 81.

Which pH?

This will depend on the type of study. For kinetic studies it is better to use a pH of 8 (or between 7.5 and 8.0) because this is the region of optimum pH of the enzyme, and small variations in pH do not affect V_{max} or $K_{0.5}$. This pH also favors the existence of $MgATP^{2-}$, and some species such as $HATP^{3-}$ and MgHATP are present at pH 7.0 but insignificant at pH 8.0. Whichever pH is used, it does not affect the cooperativity, as we have seen above. If one is interested in physiological interactions, as in studies of the regulatory protein, it is better to follow Van Schaftingen in using a physiological pH of about 7.2, but one must then be careful to prepare the buffers with great precision as small variations have significant effects on the rate, and to bear in mind that the ion $MgATP^{2-}$ is less predominant than at pH 8.0. In choosing the type of buffer one should avoid ones such as Tris that chelate Mg^{2+} ions. Triethanolamine buffers are much better.

How much K^+?

For stability reasons, and, in Assay 2, because K^+ is an activator of pyruvate kinase, it is advisable to use K^+ concentrations in the range 50-100 mM.

PURIFICATION

Hexokinase D is present in rather low concentrations in rat liver; it has been estimated that it represents about 0.01-0.1% of the total cytosolic protein in rat liver.[82] (The proportion present in islets of Langerhans is some 10-20 times smaller, as discussed in Chapter 5). As hexokinase D is also an unstable enzyme it is easy to understand why it took so long to obtain an efficient method of purification. The use of glycerol (10-20% during purification and 30% for subsequent storage), K^+ and dithiothreitol during purification proved to be crucial for stability, and the introduction of affinity chromatography led to major progress in purification. The first method that proved to be quite efficient was one developed by Ian Trayer's group,[20] which differed from the earlier ones in two principal respects: first by performing the initial step on DEAE-cellulose, introduced by González and coworkers in 1964,[83] at a lower pH of 6.5, thereby achieving a greatly improved specific activity at an early stage; and second by the inclusion of an affinity chromatography step on Sepharose-*N*-(6-aminohexanoyl)-2-amino-2-deoxy-D-glucopyranose, described in detail in reference 84.

Some variations on this procedure have been described,[31,68,85] which have allowed the purification of hexokinase D expressed in *Escherichia coli*. Establishing an efficient purification procedure was crucial for the current development of the hexokinase D field. For example, the procedure worked out by Trayer's group allowed Seitz and coworkers to purify the rat liver enzyme to homogeneity and to raise specific antibodies in sheep.[86] These antibodies were initially used to quantify the mRNA of hepatic hexokinase D by translational assay, and subsequently for screening a rat liver cDNA library that led to isolation of the first cDNA clone of the enzyme.[87]

STRUCTURE OF HEXOKINASE D, A MONOMERIC ENZYME OF 50 kDa

Hexokinase D is the only mammalian hexokinase reported to have a molecular mass as low as 50 kDa.[7,20,21] The reported values have varied widely, in the range 48-68kDa, but with homogenous preparations different methods give a molecular mass close to 50 kDa,[20] about half of the value of 96 kDa reported for the other hexokinases.[10,88] This value of about 50 kDa is now well supported by the sequence.[85]

The molecular mass studies have shown that the enzyme is a monomer, even in the presence of glucose.[20,21,64] Thus studies on polyacrylamide gels containing sodium dodecylsulfate indicated a molecular mass of about 52 kDa for hexokinase D. The same result was obtained whether or not the enzyme was reduced with 2-mercaptoethanol (as was the normal procedure) during treatment with sodium dodecylsulfate before electrophoresis. Equilibrium sedimentation measurements of the native and denatured enzyme in the analytical ultracentrifuge under different conditions gave values in the range 46.8-48.6 kDa. Gel filtration studies gave similar values, 49.5 and 48.5 kDa, this last in the presence of glucose or MgATP.

The amino acid sequence of rat liver hexokinase D has been deduced from the nucleotide sequence of a full-length cDNA,[85] and independently confirmed.[89] The sequence is 465 residues in length, with a calculated molecular mass of 51919 daltons and an isoelectric point of 4.85. The sequence is highly conserved in man, with 98% identical residues in the human and rat enzymes.[90] There is high sequence similarity between hexokinase D and the three other mammalian hexokinases. Rat hexokinases A, B and C have amino acid chain lengths of 918, 917 and 924 residues respectively.[91-94]

The liver and islet β-cell forms of hexokinase D differ by short *N*-terminal chains of 14-16 amino acids encoded by tissue-specific leader exons, whereas the remaining 450 residues in the polypeptide sequence are common to both molecules (see Figs. 5.3 and 5.4) as will be discussed in Chapter 5.

The molecular masses of the hexokinases and the similarities between their amino acid compositions have long suggested that the low-K_m hexokinases approximate to dimers of hexokinase D. This conclusion is now amply confirmed by determination of the amino acid sequences of all four isoenzymes, as we shall see below.

Similarity of the primary structure of hexokinase D to that of the other mammalian hexokinases

Comparison of the different vertebrate hexokinase isoenzymes has led to speculations about the evolution of the hexokinases. Several groups have suggested that the mammalian hexokinases evolved by duplication and fusion of an ancestral hexokinase of 50 kDa, which resembled the present-day yeast hexokinases and mammalian hexokinase D in size.[95-97]

Determination of the complete amino acid sequences of numerous hexokinases, for which extensive data are summarized in a recent review Fothergill-Gilmore and Michels,[98] has allowed a test of the predictions from the proposed evolutionary pathway, namely that extensive sequence similarity ought to exist between the *N*- and *C*-terminal regions of any low-K_m mammalian hexokinase. The *N*- and *C*-terminal halves of rat brain hexokinase A do indeed show extensive sequence similarity to each other: 49% of amino acid residues are identical, with a further 17% of conservative substitutions. Residues corresponding to the catalytically important Ser-158, Asn-210, Asp-211, Glu-269 and Glu-302 of yeast hexokinase, as well as residues proposed to be important in the binding of MgATP,[92] are all found at equivalent positions in the two halves of the enzyme. Moreover, hexokinase D resembles each half of any of the low-K_m enzymes as much as the two halves resemble each other; for example, it has 51% of identity with the *C*-terminal half of hexokinase A, 53% with the *C*-terminal half of rat hexokinase B, and 52% with the *N*-terminal half of rat hexokinase B. These may be compared with the 91% identity between human and rat hexokinase A, and provide evidence to support the hypothesis of gene duplication and fusion in the evolution of the 100kDa hexokinases from the 50kDa yeast-like enzymes, events that must have occurred long before the separation of the various mammalian lines.

Model for structure of human hexokinase D

Crystal structures are not yet available for hexokinase D, nor for any of the mammalian low-K_m hexokinases. However, the A and B isoenzymes of yeast hexokinase have been studied by x-ray crystallography.[99,100] There are two domains separated by a large cleft leading to the active site. The amino acid sequences of human hexokinase D and yeast hexokinase are about 30% identical, and residues in the active site and on the surface of the cleft separating the two lobes are highly conserved.[68] There is good reason therefore to believe that the crystallographic studies of yeast hexokinase constitute a valuable resource for modeling the mammalian hexokinase and predicting the location of functionally important amino acid residues, and it has been possible to build a molecular model of human β-cell hexokinase D based on the crystal structure of yeast hexokinase B.[101] This model suggests that the glucose-binding site in hexokinase D is formed by residues Lys-56, Asn-204, Gly-229, Asn-231, Asp-205, Glu-256, Gln-287 and Glu-290, which make interconnecting hydrogen bonds with one other and with the hydroxyls of glucose, but no interactions are predicted for the glucose ring oxygen atom at the 5 position.

The positions of these residues in relation to the exon organization of the human hexokinase D mRNA are shown in Figure 5.4. Tests of these assignments by site-directed mutagenesis of cDNA and bacterial expression have been initiated by Pilkis and collaborators, and mutational studies of several proposed glucose-binding residues (Asn-204, Asp-205 and Glu-256) are consistent with the prediction that these residues form hydrogen bonds with glucose, as mutating them caused a decrease in glucose affinity.[22] (There are, however, some problems with the way in which the kinetic data of these experiments were analyzed, as discussed earlier in this chapter). Mutations of glucose-binding residues also caused a decrease in V_{max}, which was expected from the conclusion that glucose phosphorylation is tightly coupled to the substrate-induced conformational change in the yeast enzyme during catalysis.[99] The effects of mutations in glucose-binding residues on the kinetic behavior depended on the nature of the specific mutation and was consistent with the interpretation that these residues form hydrogen bonds to the glucose hydroxyls. Only mutations in residues predicted to form hydrogen bonds with glucose, i.e. Asn-204, Glu-290 and Glu-256, had an effect on the cooperative behavior of the enzyme. The determinants for sigmoidicity and sugar binding are thus beginning to emerge. However, the location of the ATP-binding site and the identities of the residues involved in ATP binding are less clear.

PERSPECTIVES

Now that the hepatic and β-cell forms of hexokinase D can be readily expressed in *Escherichia coli* and purified to homogeneity the analysis of the effects of site-directed mutations on hexokinase D activity is greatly facilitated. Together with the availability of a molecular model for human hexokinase D, this has been extremely useful for understanding the effects of natural hexokinase D mutations in the human on enzymatic activity. The structural basis for the higher sugar affinity of hexokinases A, B and C compared with hexokinase D is still not known, even though the residues that are predicted to bind glucose are conserved in these proteins. The model for hexokinase D may allow use of site-directed mutagenesis to generate a mutant hexokinase D with a higher affinity for glucose, and to understand the cooperativity with glucose. As the amino acids involved in sugar binding are well conserved it is not easy to understand large differences in behavior, such as inhibition by excess of glucose in the case of hexokinase C and cooperativity in hexokinase D. Ultimately the crystal structure of hexokinase D will provide answers to this and other questions.

REFERENCES

1. Cárdenas ML, Rabajille E, Trayer IP et al. Cooperative interactions in hexokinase D ("glucokinase"). Kinetic and fluorescence studies. Arch Biol Med Exp 1985; 18:273-284.
2. Pollard-Knight D, Cornish-Bowden A. Mechanism of liver glucokinase. Mol Cell Biochem 1982; 44:71-80.
3. Neet KE, Tippett PS, Keenan RP. Regulatory properties of glucokinase. In: Chock PB, Huang CY, Tsou CL et al, eds., Enzyme Dynamics and Regulation. New York: Springer-Verlag, 1988:28-38.
4. González C, Ureta T, Babul J et al. Characterization of isoenzymes of adenosine triphosphate:D-hexose 6-phosphotransferase from rat liver. Biochemistry 1967; 6:460-468.
5. Parry MJ, Walker DG. Purification and properties of adenosine 5'-triphosphate-D-glucose 6-phosphotransferase from rat liver. Biochem J 1966; 99:266-274.
6. Parry MJ, Walker DG. Further properties and possible mechanism of adenosine 5'-triphosphate-D-glucose 6-phosphotransferase from rat liver. Biochem J 1967; 105:473-482.

7. Pilkis SJ. Rat hepatic glucokinase: improved purification and some properties. Arch Biochem Biophys 1972; 149:349-360.
8. Grossman SH, Dorn CG, Potter VR. The preparation and purification of pure rat liver glucokinase. J Biol Chem 1974; 249:3055-3060.
9. Salas J,Viñuela E, Salas M et al. Glucokinase of rabbit liver. Purification and properties. J Biol Chem 1965; 240:1014-1018.
10. Niemeyer H, Ureta T, Clark-Turri L. Adaptive character of liver glucokinase. Mol Cell Biochem 1975; 6:109-126.
11. Niemeyer H, Cárdenas ML, Rabajille E et al. Sigmoidal kinetics of glucokinase. Enzyme 1975; 20:321-333.
12. Storer AC, Cornish-Bowden A. Kinetics of rat liver glucokinase. Co-operative interactions with glucose at physiologically significant concentrations. Biochem J 1976; 159:7-14.
13. Storer AC, Cornish-Bowden A. Kinetic evidence for a "mnemonical" mechanism for rat liver glucokinase. Biochem J 1977; 165:61-69.
14. Kelkar SM, Nadkarni GB. Regulatory responses of rat liver glucokinase. Indian J Biochem Biophys 1978; 15:95-100.
15. Tippett PS, Neet KE. Specific inhibition of glucokinase by long chain acyl coenzymes A below the critical micelle concentration. J Biol Chem 1982; 257:12839-12845.
16. Tippett PS, Neet KE. An allosteric model for the inhibition of glucokinase by long chain acyl coenzyme A. J Biol Chem 1982; 257:12846-12852.
17. Ureta T, Radojkovic J, Lagos R et al. Phylogenetic and ontogenetic studies of glucose phosphorylating isoenzymes of vertebrates. Arch Biol Med Exp 1979; 12:587-609.
18. Cárdenas ML, Rabajille E, Niemeyer H. Kinetic cooperativity of glucokinase with glucose. Arch Biol Med Exp 1979; 12:571-580.
19. Meglasson MD, Burch PT, Berner DK et al. Chromatographic resolution and kinetic characterization of glucokinase from islets of Langerhans. Proc Natl Acad Sci USA 1983; 80:85-89.
20. Holroyde MJ, Allen MB, Storer AC et al. The purification in high yield and characterization of rat hepatic glucokinase. Biochem J 1976; 153 :363-373.
21. Cárdenas ML, Rabajille E, Niemeyer H. Maintenance of the monomeric structure of glucokinase under reacting conditions. Arch Biochem Biophys 1978; 190:142-148.
22. Pilkis SJ, Weber IT, Harrison RW et al. Glucokinase: structural analysis of a protein involved in susceptibility to diabetes. J Biol Chem 1994; 269:21925-21928.
23. Radojkovic J, Slebe JC, Ureta T. A radioassay for the glucokinase reaction. Arch Biol Med Exper 1978; 11:73-76.
24. Bontemps F, Hue L, Hers H-G. Phosphorylation of glucose in isolated rat hepatocytes. Sigmoidal kinetics explained by the activity of glucokinase alone. Biochem J 1978; 174:603-611.
25. Clark DG, Rognstad R, Katz J. Isotopic evidence for futile cycles in liver cells. Biochem Biophys Res Comm 1973; 54:1141-1148.

26. Katz J, Wals PA, Golden S et al. Recycling of glucose by rat hepatocytes. Eur J Biochem 1975; 60:91-101.
27. Newsholme EA, Start C. In: Regulation in Metabolism. London: Wiley, 1973:267-270.
28. Hue L, Hers H-G. Utile and futile cycles in the liver. Biochem Biophys Res Comm 1974; 58:540-548.
29. Barzilai N, Rossetti L. Role of hexokinase D and glucose-6-phosphatase in the acute and chronic regulation of hepatic glucose fluxes by insulin. J Biol Chem 1993; 268:25019-25025.
30. Cornish-Bowden A. In: Analysis of Enzyme Kinetic Data. Oxford: Oxford University Press, 1995:57-59.
31. Lange AJ, Xu LZ, Van Poelwijk F et al. Expression and site-directed mutagenesis of hepatic glucokinase. Biochem J 1991; 277:159-163.
32. Clark-Turri L, Peñaranda J, Rabajille E et al. Immunochemical titration of liver glucokinase from normal, fasted and diabetic rats. FEBS Lett 1974; 41:342-344.
33. Pilkis SJ, Hansen RJ, Krahl ME. Hexose-ATP phosphotransferases: comparative aspects. Comp Biochem Physiol 1968; 25:903-912.
34. Ureta T. The allosteric regulation of hexokinase C from amphibian liver. J Biol Chem 1976; 251:5035-5042.
35. Radojkovic J, Ureta T. Hexokinase isoenzymes from the Novikoff hepatoma. Purification, kinetic and structural characterization, with emphasis on hexokinase C. Biochem J 1987; 242:895-903.
36. Meunier J-C, Buc J, Navarro A et al. Regulatory behavior of monomeric enzymes. 2. A wheat-germ hexokinase as a mnemonical enzyme. Eur J Biochem 1974; 49:209-223.
37. Cárdenas ML, Rabajille E, Niemeyer H. Fructose: a good substrate for rat-liver "glucokinase" (hexokinase D). Biochem J 1984; 222:363-370.
38. Niemeyer H, Monasterio O, Cárdenas ML et al. Kinetic mechanism of the phosphorylation of 2-deoxyglucose by rat liver glucokinase. Fed Proc 1978; 37:1421.
39. Monasterio O. Mecanismo cinético de la glucoquinasa. Doctoral thesis, University of Chile, 1980.
40. Cárdenas ML, Rabajille E, Niemeyer, H. Suppression of kinetic cooperativity of hexokinase D ("glucokinase") by competitive inhibitors. Eur J Biochem 1984; 145:163-171.
41. Randle PJ. Glucokinase and candidate genes for type 2 (non-insulin-dependent) diabetes mellitus. Diabetologia 1993; 36:269-275.
42. Meglasson MD, Schinco M, Matschinsky FM. Mannose phosphorylation by glucokinase from liver and transplantable insulinoma: cooperativity and discrimination of anomers. Diabetes 1983; 32:1146-1151.
43. Pigman W, Isbell HS. Mutarotation of sugars in solution: Part I. History, basic kinetics, and composition of sugar solutions. Adv. Carbohyd. Chem. 1968; 23:57-77.
44. Meglasson MD, Matschinsky FM. Discrimination of glucose anomers by glucokinase from liver and transplantable insulinoma. J Biol Chem 1983; 258:6705-6708.

45. Miwa J, Inagaki K, Okuda J. Preference of glucokinase for the α-anomer of hexose: relation to α-anomer preference in hexose-induced insulin release by pancreatic islets. Biochem Int 1983; 7:449-454.
46. Ferdinand W. The interpretation of non-hyperbolic rate curves for two-substrate enzymes. A possible mechanism for phosphofructokinase. Biochem J 1966; 98:278-283.
47. Ureta T, Radojkovic J, Díaz N et al. Comparative studies on glucose phosphorylating isoenzymes of vertebrates. Identification and characterization of amphibian liver hexokinases. Arch Biochem Biophys 1978; 186:235-247.
48. Vandercammen A, Van Schaftingen E. Competitive inhibition of liver glucokinase by its regulatory protein. Eur J Biochem 1991; 200:545-551.
49. Walker DG. The nature and function of hexokinases in animal tissues. Essays Biochem 1966; 2:33-37.
50. Pollard-Knight D, Cornish-Bowden A. Kinetics of hexokinase D ("glucokinase") with inosine triphosphate as phosphate donor. Biochem J 1987; 245:625-629.
51. Pollard-Knight DV, Cornish-Bowden A. Solvent isotope effects on the glucokinase reaction. Negative co-operativity and a large inverse isotope effect in $^{2}H_2O$. Eur J Biochem 1984; 141:157-163.
52. Dixon M. The effect of pH on the affinities of enzymes for substrates and inhibitors. Biochem J 1953; 55:161-170.
53. Mansour TE, Ahlfors CE. Studies on heart phosphofructokinase. Some kinetic and physical properties of the crystalline enzyme. J Biol Chem 1968; 243:2523-2533.
54. Rozengurt E, Jiménez de Asúa L, Carminatti H. Some kinetic properties of liver pyruvate kinase (type L). J Biol Chem 1969; 244:3142-3147.
55. Rübsamen H, Khandker R, Witzel H. Sigmoid kinetics of the monomeric ribonuclease I due to ligand-induced shifts of conformation equilibria. Hoppe-Seyler's Z Physiol Chem 1974; 355:687-708.
56. Walker EJ, Ralston GB, Darvey IG. Further evidence for an allosteric model for ribonuclease. Biochem J 1976; 153:329-337.
57. Walker EJ, Ralston GB, Darvey IG. The nature of the allosteric interactions of ribonuclease and its ligands. Biochem J 1978; 173:1-4.
58. Cárdenas ML. Glucoquinasa, una enzima monomérica con cinética cooperativa. Doctoral thesis, University of Chile, 1982.
59. Colowick SP. The hexokinases. In: Boyer PD, ed. The Enzymes. 3rd ed. New York: Academic Press, 1973; 9:1-48.
60. Monasterio O, Orellana O, Heberlein U et al. Protective effect of various ligands against glucokinase inactivation by 5,5'-dithiobis-2-nitrobenzoic acid. Arch Biol Med Exper 1981; 14:139-142.
61. Niemeyer H, Heberlein U, Monasterio O. Dissociation between catalytic activity and susceptibility to inactivation by 5,5'-dithiobis-2-nitrobenzoic acid of rat glucokinase. Arch Biol Med Exper 1981; 14:177-182.
62. Frieden C. Kinetic aspects of regulation of metabolic processes. The hysteretic enzyme concept. J Biol Chem 1970; 245:5788-5799.
63. Cohen LA. Group-specific reagents in protein chemistry. Ann. Rev.

Biochem. 1968; 37:695-726.

64. Maccioni R, Babul J. Purification and characterization of dog liver glucokinase. Arch Biol Med Exper 1980; 13:271-286.
65. Kristensen Ø. Energization of alanine transport in isolated rat hepatocytes. Electrogenic Na^+-alanine co-transport leading to increased K^+ permeability. J Biol Chem 1980; 255:5236-5243.
66. Meijer AJ, Baquet A, Gustafson L et al. Mechanism of activation of liver glycogen synthase by swelling. J Biol Chem 1992; 267:5823-5828.
67. Cárdenas ML, Cornish-Bowden A. Rounding error, an unexpected fault in the output from a recording spectrophotometer: implications for model discrimination. Biochem J 1993; 292:37-40.
68. Gidh-Jain M, Takeda J, Xu LZ et al. Glucokinase mutations associated with non-insulin-dependent (type 2) diabetes mellitus have decreased enzymatic activity: Implications for structure/function relationships. Proc Natl Acad Sci USA 1993; 90:1932-1936.
69. Walker DG, Parry MJ. Glucokinase. I. Liver. Methods Enzymol. 1966; 9:381-388.
70. Storer AC, Cornish-Bowden A. The kinetics of coupled enzyme reactions. Applications to the assay of glucokinase, with glucose 6-phosphate dehydrogenase as coupling enzyme. Biochem J 1974; 141:205-209.
71. Slein MW, Cori GT, Cori CF. A comparative study of hexokinase from yeast and animal tissues. J Biol Chem 1950; 186:763-780.
72. Kornberg A, Pricer WE. Enzymatic phosphorylation of adenosine and 2,6-diaminopurine riboside. J Biol Chem 1951; 193:481-495.
73. Greiling H, Kisters R. Studies on substrate specificity of glucose 6-phosphate dehydrogenase. Hoppe-Seyler's Z Physiol Chem 1965; 341:172-184.
74. Anstall HB, Trujillo JM. Glucose 6-phosphate dehydrogenase. Its purification and properties. Amer J Clin Pathol 1967; 47:296-302.
75. Broyles RH, Strittmayer CF. Hexose monophosphate shunt dehydrogenases in the sea urchin and the frog: comparison of some functional properties of the enzymes in vitro. Comp Biochem Physiol 1977; 57B:249-255.
76. Tokarev YN, Chernyak NB, Batischev AI et al. Study of electrophoretic and kinetic properties of G6PD from erythrocytes in cases of inherited enzyme deficiency. Nouvel Rev Fr Hematol 1978; 20:558-564.
77. Stanley JC, Dohm GL, McManus BS et al. Activities of glucokinase and hexokinase in mammalian and avian livers. Biochem J 1984; 224:667-671.
78. Bedoya FJ, Meglasson MD, Wilson JM et al. Radiometric oil well assay for glucokinase in microscopic structures. Anal Biochem 1985; 144:504-513.
79. Storer AC, Cornish-Bowden A. Concentration of $MgATP^{2-}$ and other ions in solution. Calculation of the true concentrations of species present in mixtures of associating ions. Biochem J 1976; 159:1-5.
80. MacFarlane N, Ainsworth S. A kinetic study of baker's-yeast pyruvate kinase activated by fructose 1,6-diphosphate. Biochem J 1972; 129:1035-1047.
81. Cornish-Bowden A. In: Fundamentals of Enzyme Kinetics. 2nd ed. London: Portland Press, 1995:69-72.
82. Iynedjian PB. Mammalian glucokinase and its gene. Biochem J 1993; 293:1-13.

83. González C, Ureta T, Sánchez R et al. Multiple molecular forms of ATP:hexose 6-phosphotransferase from rat liver. Biochem Biophys Res Commun 1964; 16:347-352.
84. Holroyde MJ, Chesher JME, Trayer IP et al. Studies on the use of Sepharose-*N*-(6-aminohexanoyl)-2-amino-2-deoxy-D-glucopyranose for the large-scale purification of hepatic glucokinase. Biochem J 1976; 153:351-361.
85. Andreone TL, Printz RL, Pilkis SJ et al. The amino acid sequence of rat liver glucokinase deduced from cloned cDNA. J Biol Chem 1989; 264:363-369.
86. Minderop RH, Hoeppner W, Seitz HJ. Regulation of hepatic glucokinase gene expression. Role of carbohydrates, and glucocorticoid and thyroid hormones. Eur J Biochem 1987; 184:181-187.
87. Iynedjian PB, Ucla C, Mach B. Molecular cloning of glucokinase cDNA. Developmental and dietary regulation of glucokinase mRNA in rat liver. J Biol Chem 1987; 262:6032-6038.
88. Wright CL, Warsy AS, Holroyde MJ et al. Purification of the hexokinases by affinity chromatography on Sepharose-*N*-aminoacylglucosamine derivatives. Design of affinity matrices from the free solution kinetics. Biochem J 1978; 175:125-135.
89. Hayzer DJ, Iynedjian PB. Alternative splicing of glucokinase mRNA in rat liver. Biochem J 1990; 270:261-263.
90. Tanizawa Y, Koranyi LI, Welling CM et al. Human liver glucokinase gene: cloning and sequence determination of two alternatively spliced cDNAs. Proc Natl Acad Sci USA 1991; 88:7294-7297.
91. Schwab DA, Wilson JE. The complete amino acid sequence of the catalytic domain of rat brain hexokinase, deduced from the cloned cDNA. J Biol Chem 1988; 263:3220-32[illegible].
92. Schwab DA, Wilson JE. Complete amino acid sequence of rat brain hexokinase, deduced from the cloned cDNA, and proposed structure of a mammalian hexokinase. Proc Natl Acad Sci USA 1989; 86:2563-2567.
93. Thelen AP, Wilson JE. Complete amino acid sequence of the type II isozyme of rat hexokinase, deduced from the cloned cDNA: comparison with a hexokinase from Novikoff ascites tumor. Arch Biochem Biophys 1991; 286:645-651.
94. Schwab DA, Wilson JE. Complete amino acid sequence of the type III isoenzyme of rat hexokinase, deduced from the cloned cDNA. Arch Biochem Biophys 1991; 285:365-370.
95. Holroyde MJ, Trayer IP. Purification and properties of rat skeletal muscle hexokinase. FEBS Lett 1976; 62:215-219.
96. Ureta T. The comparative isozymology of vertebrate hexokinases. Comp Biochem Physiol 1982; 71B:549-555.
97. Ureta T, Medina C, Preller A. The evolution of hexokinases. Arch Biol Med Exp 1987; 20:343-357.
98. Fothergill-Gilmore LA, Michels PAM. Evolution of glycolysis. Prog Biophys Molec Biol 1993; 59:105-235.

99. Anderson CM, Zucker FH, Steitz TA. Space-filling models of kinase clefts and conformation changes. Science 1979; 204:375-380.
100. Bennett WS, Steitz TA. Structure of a complex between yeast hexokinase A and glucose. II. Detailed comparisons of conformation and active site configuration with the native hexokinase B monomer and dimer. J Mol Biol 1980; 140:211-230.
101. St. Charles R, Harrison RW, Bell GI et al. Molecular model of the human β-cell glucokinase built by analogy to the crystal structure of yeast hexokinase B. Diabetes 1994; 43:784-791.

CHAPTER 4

Tissue and Cellular Distribution of Hexokinase D

Hexokinase D has been known to exist in hepatocytes for many years, and more recently its presence in pancreatic β-cells has become established. However, during the whole period there have been recurrent reports of hexokinase D in other tissues and cells. All of those that have been thoroughly studied have proved to be artifacts of the presence of *N*-acetylglucosamine kinase, an enzyme that is far more specific for *N*-acetylglucosamine than for glucose, but which shares certain properties with hexokinase D that encourage confusion between the two if *N*-acetylglucosamine is not tested as a substrate. The existence within the liver not only of hepatocytes but also of several other types of cells needs to be considered in assessing the roles of the hexokinases. All four isoenzymes are present in hepatocytes, but the glucose phosphorylating capacity of these cells is overwhelmingly due to hexokinase D, which is absent from other liver cells. Hexokinase D activity is highest in hepatocytes in the perivenous region of the liver, whereas glucose 6-phosphatase activity is highest in hepatocytes in the periportal region. Cycling between glucose and glucose 6-phosphate may therefore occur to a much smaller extent than the activities of the two enzymes averaged over the whole population of hepatocytes would suggest.

EXTRAHEPATIC HEXOKINASE D: MYTH OR REALITY?

The hexokinase reaction is ubiquitous.[1] However, not every isoenzyme is present in each type of cell. This is especially true of hexokinase D, which has a very restricted localization and is the characteristic hexokinase of hepatocytes. Several research groups have reported the presence of a high-K_m hexokinase in tissues other than adult liver, but until this moment the only clearly established extrahepatic hexokinase D is that of the pancreas, described in Chapter 5. It is true that a high-K_m glucose phosphorylating activity is present in several organs besides liver, but the enzyme responsible has been clearly identified as *N*-acetylglucosamine kinase.[2-4]

N-ACETYLGLUCOSAMINE KINASE MISTAKENLY IDENTIFIED AS HEXOKINASE D

N-Acetylglucosamine kinase (ATP:acetamide-2-deoxy-D-glucose-6-phosphotransferase, EC 2.7.1.59), has *N*-acetylglucosamine as its specific substrate, but it can also phosphorylate glucose with very low affinity. Because of this property, it has been mistakenly identified as "glucokinase" (in the sense of hexokinase D) in several tissues, e.g. intestinal mucosa, human placenta, rat kidney, liver from new-born or fetal rats. The criteria for making this identification have been varied. In some cases hexokinase D activity has been sought by measuring glucose phosphorylation in crude extracts at a high and a low glucose concentration, usually 100 mM and 0.5 mM respectively, and a ratio of rates at the two concentrations of at least 1.5 is considered to indicate hexokinase D activity. In others it has been the finding of curved Lineweaver-Burk plots (or related plots) for glucose phosphorylation, from which a high K_m for glucose has been calculated. Finally, the separation by electrophoresis or chromatography of a protein fraction with a mobility very similar to that of the liver enzyme and catalyzing glucose phosphorylation only at high sugar concentration has also been considered as a sufficient proof of the presence of hexokinase D. As discussed below, all of these criteria may be necessary as *preliminary* diagnostic tools, but they are not sufficient to prove the presence of hexokinase D, and must be checked by alternative procedures.

Several groups showed that there is indeed a glucose-phosphorylating activity with a high half-saturation concentration in numerous different tissues, including spleen, placenta, neonatal liver, mammary gland, kidney, intestinal mucosa, lung and heart muscle.[2-4] However, further examination of the properties of these putative hexokinases D showed with certainty that they correspond to *N*-acetylglucosamine kinase, and cast serious doubts on the presence of hexokinase D in nearly all extrahepatic cells. To this moment, the only definite extrahepatic hexokinase D is the one present in the β-cells of pancreatic islets and insulinomas.[5-9] Not only is this enzyme clearly different from *N*-acetylglucosamine kinase, but, as we shall see in Chapter 5, it has now been established that it is coded by the same gene as the hepatic hexokinase D, explaining the great similarity in structural and kinetic properties between the two enzymes.

The suggestion of "glucokinase" in tissues other than adult liver has always raised interest because it would indicate the possible operation therein of processes to regulate glucose metabolism. Thus, the possible presence of a high-K_m hexokinase in cattle spleen appeared quite exciting, as in these animals not even the liver has hexokinase D, as we saw in Chapter 1. However, when the enzyme was purified to homogeneity (only one band in polyacrylamide electrophoresis stained with Coomassie blue) it proved able to phosphorylate not only glucose, mannose and 2-deoxyglucose, like hexokinase D, but also *N*-acetylglucosamine, which is not a substrate of hexokinase D. The apparent half-saturation concentrations of this enzyme with *N*-acetylglucosamine and glucose were 0.066 mM and 370 mM respectively, indicating that *N*-acetylglucosamine is by far the preferred substrate; glucose is a very poor substrate indeed![2] The kinetic behavior with glucose as a substrate was definitively nonhyperbolic, suggesting negative cooperativity, and the reported K_m value was found by extrapolating rate measurements in the range 150-600 mM. (Recall that hexokinase D shows positive cooperativity instead, as described in Chapter 3). A similarly high half-saturation concentration was observed in intestinal mucosa.[4] Furthermore, an immune serum against cattle spleen *N*-acetylglucosamine kinase did not cross-react with purified hexokinases A, B, C and D. It could, however, remove the supposed hexokinases D from extracts of rat kidney, newborn rat liver and one of the two electrophoretic bands previously attributed to liver hexokinase D, namely the one now known to be *N*-acetylglucosamine kinase; it had no effect on the genuine hexokinase D.[2]

It is not difficult to understand why so many investigators have mistakenly identified hexokinase D in extrahepatic tissues, because several properties of *N*-acetylglucosamine kinase do resemble those of liver hexokinase D; both enzymes display high half-saturation concentrations for glucose, both show lower activity with fructose at 100 mM, and both can phosphorylate other sugars like mannose and 2-deoxyglucose. Moreover, they have similar physicochemical properties, such as electrophoretic and chromatographic mobility, and molecular mass (about 55 kDa), and they tend to copurify even in steps of affinity chromatography.

Complex kinetics of *N*-acetylglucosamine kinase with glucose as substrate

The kinetic behavior of *N*-acetylglucosamine kinase with glucose as substrate requires some comment, to understand why authors who thought that it might be a hexokinase D obtained a K_m value for glucose of the order of 10 mM, whereas others reported much higher or much lower values. The saturation function of *N*-acetylglucosamine kinase with glucose is non-Michaelian, generally mimicking negative cooperativity. Thus double-reciprocal plots show downward curvature, which does not allow estimation of a K_m value, and if the range of glucose concentrations is narrow the curvature may not be perceived and the results will be misleading. This point is well brought out by the work on the enzyme from intestinal mucosa,[4] illustrated in Figure 4.1. After rates had been measured in the range 4-400 mM glucose, three subsets of experimental data were selected for drawing straight lines on the double-reciprocal plot. Each subset of data fitted a straight line reasonably well, and depending on which set of points was considered the value of K_m for glucose could be estimated as 1, 20 or 280 mM, as shown in the insets to Figure 4.1. Similar kinetic behavior has been observed with pure enzyme from cattle spleen, and with semipurified enzymes from human placenta and rat kidney. It follows that in any of these cases if the curvature of the plot is overlooked the "K_m" reported is likely to be arbitrarily dependent on the particular range of glucose concentrations used in the plot, so that it is perfectly possible to obtain values ranging over a factor of 300 or more from the same enzyme measured under the same conditions.

This variation in the "K_m" obtained is a very important point to understand, because in usual practice the range of concentrations used for estimating a K_m or $K_{0.5}$ value is a function of the parameter value expected, and consequently an experimenter looking for a hexokinase D activity is likely to expect a $K_{0.5}$ value of about 10 mM and hence to

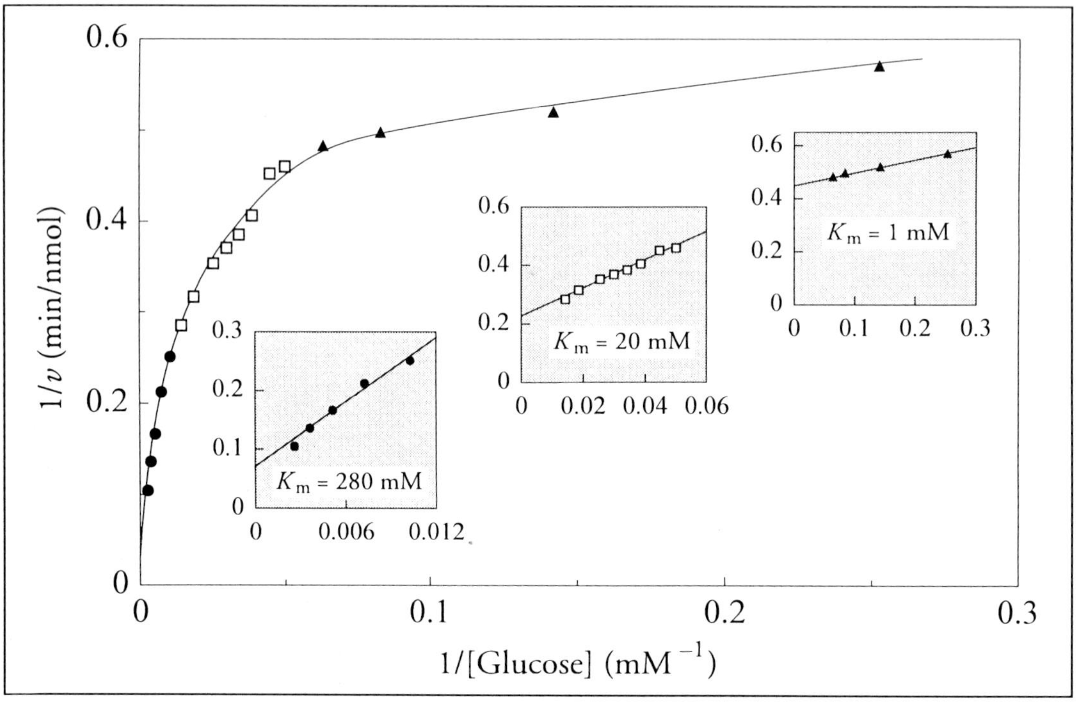

Fig. 4.1. Kinetics of N-acetylglucosamine kinase. As the double-reciprocal plot shows pronounced curvature over a wide range of concentrations, selection of a few points spanning a narrow range of concentrations and treating the resulting plot as if it were straight can yield "K_m" values that may differ by a factor of 300 or more depending on the concentration range used. The figure is redrawn from data in Figure 7 of Arch Biochem Biophys 1984; 229:237-245.

choose 5-10 concentrations in the range 2-50 mM; such an experimenter will obtain a K_m much closer to 10 mM than the actual $K_{0.5}$ of *N*-acetylglucosamine kinase, 300-400 mM, and will naturally conclude, therefore, that the activity is due to hexokinase D and not to *N*-acetylglucosamine kinase. This has become an especially serious problem with the widespread availability of user-friendly computer software that allows kinetic parameters to be estimated very easily without having to pay much attention to the appearance of the data on a graph.

A convenient diagnostic test, of course, is provided by the substrate specificity, as hexokinase D does not phosphorylate *N*-acetylglucosamine, which is an inhibitor, nor does it phosphorylate glucosamine. So any report of an extrahepatic hexokinase D cannot be taken seriously unless *N*-acetylglucosamine kinase has been explicitly ruled out. To my knowledge the only extrahepatic hexokinase D that has survived this test is the enzyme of pancreatic β-cells.

Tissue-specific gene expression

The narrow tissue distribution of hexokinase D has been confirmed with very sensitive techniques such as the immunoblotting method. Using an antibody raised against liver hexokinase D, for example, this procedure has revealed a single band corresponding to a polypeptide with an apparent molecular mass of 56 kDa in the cytosolic fraction of liver and islet homogenates, but not in brain, spleen, intestinal mucosa, exocrine pancreas, kidney or white adipose tissue,[10] in agreement with previous reports.[2-4]

As could be expected from the distribution of the enzyme, hexokinase D mRNA has been detected by Northern blotting of poly(A)-containing RNA only, in the liver and Islets of Langerhans of glucose-fed rats.[11] As presented in Chapter 5, this mRNA has different lengths in rat liver and β-cells; in the pancreas there is a longer 5' untranslated region, encoded by the exon 1β, which is specific to β-cells. It is true that hexokinase D mRNA has also been found in other tissues, but this finding per se does not prove that active hexokinase D is present in those cells. For example, the long version of hexokinase D mRNA has been detected in the rat pituitary gland (anterior lobe) and in the corticotroph pituitary cell line AtT-20, indicating transcription of a β-cell type of promoter.[12] However, because of the preferential use of alternative RNA splice acceptor sites, described in Chapter 5, that disrupt the reading frame, hexokinase D activity was undetectable in anterior pituitary gland despite the presence of an immunoreactive protein of about 52 kDa.[12] The results were confirmed by the finding that, although hexokinase D mRNA is expressed in the pituitary from the upstream promoter, hexokinase D activity was not detectable.[13] Amplification and sequence analysis showed that alternative splicing events had occurred, either of which could, by the effects that they produce, explain the lack of hexokinase D activity.

Is there hexokinase D in brain or gut?

Very recently, the presence of hexokinase D mRNA has been reported in some neuroendocrine cells in the brain and gut,[14] and the authors also claim to have identified an active hexokinase D in these cells. This latter point is less clear in my view because, as argued above, the presence of mRNA does not guarantee the presence of active hexokinase D, and moreover the existence of a high-K_m hexokinase activity with chromatographic profile similar to that of hexokinase D does not prove that it is hexokinase D. The authors think that they have ruled out *N*-acetylglucosamine kinase because they obtained a K_m of about 10-13 mM, instead of 370 mM reported for *N*-acetylglucosamine kinase. However, with the narrow range of glucose concentrations used (1.4-21 mM), they could not have obtained a K_m of 370 mM even if it were *N*-acetylglucosamine kinase, because of the non-Michaelian kinetics of this enzyme, as discussed above. In other words, if the enzyme in these cells were *N*-acetylglucosamine kinase and it behaved like the corresponding enzyme in spleen, placenta, liver, intestinal mucosa and kidney the experiment as described could perfectly well have yielded the apparent K_m of 10 mM reported.

Consequently, although it is certainly possible that there is a hexokinase D in the brain and gut, which would make physiological sense from the point of view of the response to glucose, this cannot be asserted with any confidence at this moment. We must wait until the possibility of *N*-acetylglucosamine kinase is explicitly ruled out by establishing that the hexokinase activity with a high K_m for glucose cannot phosphorylate *N*-acetylglucosamine.

If one considers that the low affinity and sigmoidal saturation curve of hexokinase D for glucose allow this isoenzyme to respond sensitively to changes in glycemia by varying its catalytic activity, it is not surprising that it should be confined to two types of cells that play an important role in controlling glycemia, the hepatocytes and the β-cells of the islets of Langerhans. An important point to add is that these two types of cells have a very active glucose transporter (GLUT2) that allows the intracellular glucose concentration to be, most of the time, almost the same as that of the plasma.[15]

HEXOKINASE D DISTRIBUTION IN LIVER

Is hexokinase D the only hexokinase present in hepatocytes?

As we have seen above, active hexokinase D appears to be confined to two tissues, liver and pancreas. As the liver contains several types of cells,[16] it was a challenge to know the cellular location of each glucose-phosphorylating enzyme. Relatively early in the investigation of hexokinase D several authors postulated that it was present exclusively in hepatocytes, and that the high-affinity hexokinases were confined to nonparenchymal liver cells, i.e. mainly Kupffer and endothelial cells.[17,18,19] The situation was far from being clear-cut, however, as other investigators had described the presence of one or two high-affinity hexokinases activities in hepatocytes.[20,21,22] Furthermore, the presence in liver of *N*-acetylglucosamine kinase, which also phosphorylates glucose with low affinity and has sometimes been mistakenly identified as hexokinase D, as we have seen, contributed to complicate things even more.

Uncertainty on this subject persisted for a while because the methods necessary for establishing the cellular distribution with certainty had yet to be perfected, and serious criticisms could be made about the methods used, whether in isolating the cells, the quality of the isolated cells, the estimation of the amounts of contaminant cells, or the detection of the enzyme activities. One must remember that the methods used at first for separating hepatocytes yielded cells with low viability and with major losses of cytoplasmic enzymes, as reviewed by Seglen.[23]

I thought that it was important to resolve this controversy, because for any complete discussion of the physiological role of hexokinase D in the liver it was important to establish whether it was the only hexokinase isoenzyme present in the hepatocyte. With this in mind we prepared hepatocytes and nonparenchymal cells with high viability and almost no cross-contamination, using in situ perfusion of the liver with collagenase, and an enriched cell-culture medium in all steps of cell isolation.[24] This precaution proved to be crucial for detecting isoenzyme B, which was not detected when Krebs-Henseleit bicarbonate buffer was used for perfusion. We determined each hexokinase isoenzyme separately by carrying out a partial purification on small DEAE-cellulose columns and using radioassay,[25] an analytical method of great sensitivity, to detect minor amounts of high-affinity hexokinases. Figure 4.2 shows typical chromatographic profiles of hexokinase isoenzymes from high-speed supernatant fluids of hepatocytes and nonparenchymal cells isolated from the same liver.[24] It may be seen that hexokinase D occurs only in the hepatocytes, where it accounts for about 95% of the total hexokinase activity at high glucose concentrations, and that cytosol from isolated hepatocytes contains hexokinases A, B and C as well as hexokinase D. In contrast, isolated nonparenchymal cells showed only hexokinases A, B and C (Fig. 4.2b). Only hexokinase A is found associated to the particulate fraction, in isolated hepatocytes or in whole liver.[24] However, as will be discussed at the end of the chapter, hexokinase D appears to exist in a bound state in cultured hepatocytes.

N-Acetylglucosamine kinase activity was measured by a specific radioassay using *N*-acetyl-[1-^{14}C]glucosamine as substrate,[24] and was found as a single enzyme in both hepatocytes and nonparenchymal cells, with higher activities in the former. No activity not retained by DEAE-cellulose could be detected. As we shall see below, hepatocytes are metabolically heterogeneous with respect to their content of

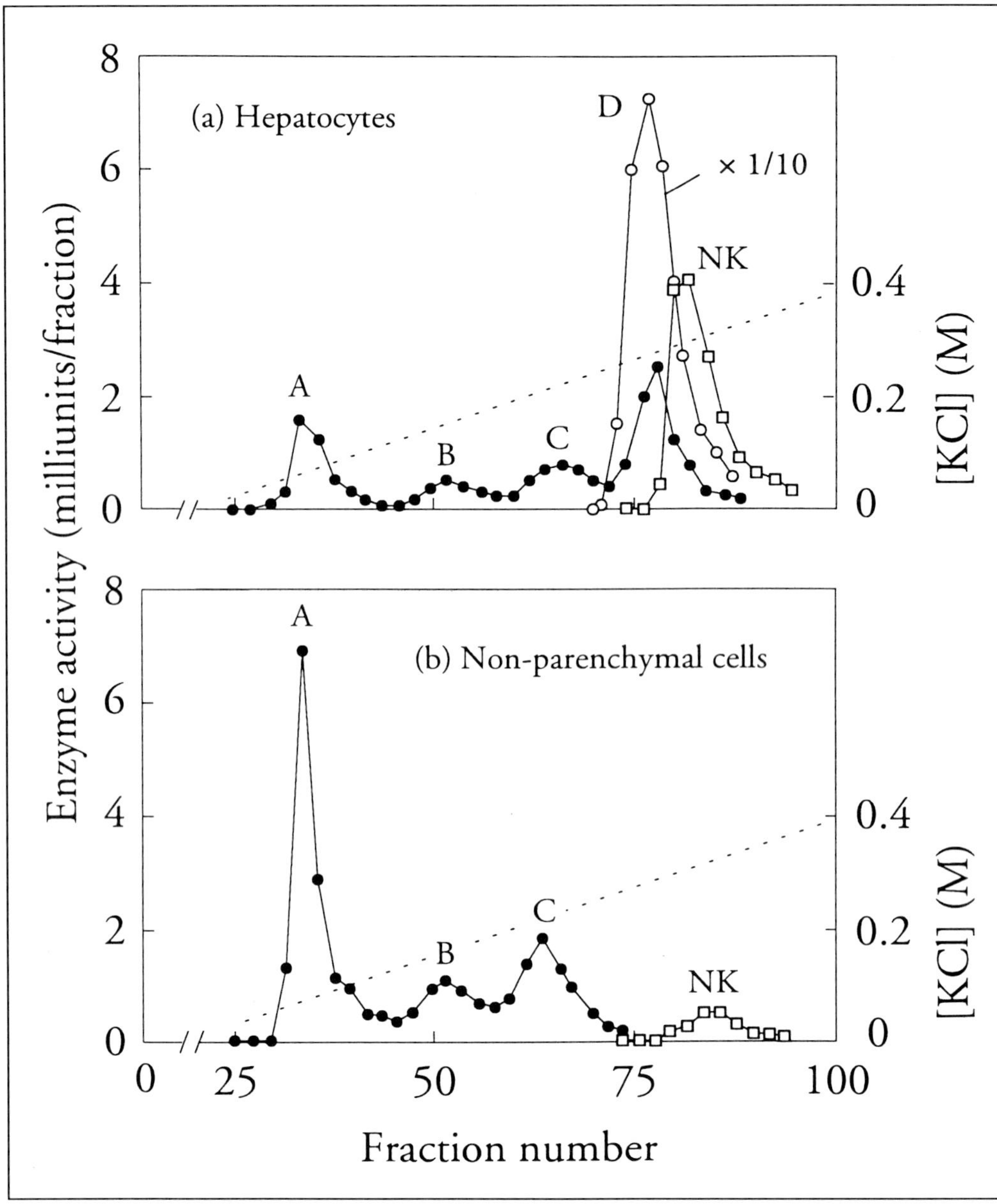

Fig. 4.2. Cellular distribution of glucose phosphorylation activity in rat liver. (a) All hexokinase isoenzymes (curves labelled A, B, C and D) are present in hepatocytes, as well as N-acetylglucosamine kinase (❑, curves NK), but hexokinase D (for which the values plotted are 1/10 those actually observed) accounts for about 95% of the total. (b) Hexokinase D is not detectable in non-parenchymal cells but all of the other three isoenzymes are present, as well as N-acetylglucosamine kinase. Hexokinase D activity was measured spectrophotometrically at 100 mM glucose (○); the other hexokinase isoenzymes were measured by radioassay at 0.5 mM glucose (●). The enzymes were eluted from DEAE-cellulose columns using the KCl gradient indicated. Reproduced with permission from Biochem J 1984; 221:303-309.

hexokinase D and other enzymes; the results in Figure 4.2a thus obviously refer to an average cell.

The uncertainty about the presence or absence of high-affinity hexokinases in hepatocytes had persisted because of methodological limitations that still apply to some degree nowadays and which are consequently worthwhile to discuss. The standard method for distinguishing the three high-affinity hexokinases from hexokinase D is to assay them spectrophotometrically (or radiometrically) at two glucose concentrations, 0.5 mM and 100 mM, as proposed in 1963.[26] At the low concentration hexokinase D activity is virtually undetectable and so this measures the high-affinity activity, whereas at the high concentration the total hexokinase activity is measured, so the hexokinase D activity can be found by subtraction. (In principle the inhibition of hexokinase C by excess glucose complicates this analysis, but this isoenzyme is present in such small quantities in all normal tissues, if it is detectable at all, that the problem can be ignored.) However, this procedure does not permit identification of the individual isoenzymes present and provides only an approximate measure of the total of high-affinity hexokinases; in some circumstances it does not even establish whether any high-affinity hexokinases are present at all. For example, when the sample contains a high proportion of hexokinase D, as in normal hepatocytes, its contribution to the activity at 0.5 mM glucose is not negligible, and levels of high-affinity hexokinases of about 5% would be difficult or impossible to detect.

The opposite case of a high activity of high-affinity hexokinases, as in the pancreas,[7] represents less of a problem, because use of a radioassay with [^{14}C]glucose allows the high-affinity hexokinases to be inhibited by addition of glucose 6-phosphate. Of course, the problem remains that the individual high-affinity isoenzymes are not distinguished from one another.

It is consequently essential to use a method that allows rapid separation of the hexokinase isoenzymes combined with an analytical method of great sensitivity. The different isoenzymes can then be assessed separately. In addition, the great instability of hexokinases B and D means that the use of protecting agents such as glycerol, dithiothreitol and KCl, as well as an enriched medium during perfusion and isolation, become crucial, so as to achieve a satisfactory recovery of enzyme activity, of the order of 90%. The problem of leaking and inactivation is a serious one and it is important to be aware of it, especially now that present techniques allow study of mutant enzymes that may well be less stable than the normal forms.

The residual phosphorylating activity after fasting or diabetes: is it *N*-acetylglucosamine kinase or hexokinase D?

As we saw in Chapter 1, the activity of hexokinase D in the liver is markedly affected by diet and hormones. On the other hand the total activity of high-affinity hexokinases appears to be constant, although some authors have described variations in certain conditions. For example, Machiya and Hoyosa reported decreased hexokinase B activity in the livers of diabetic rats;[27] Pilkis reported an increase in the total high-affinity activity in the livers of hypophysectomized rats treated with thyroxine;[28] and Walker's group observed a similar increase in the livers of neonatal rats treated with triiodothyronine.[29]

The realization that *N*-acetylglucosamine kinase is easy to misidentify as hexokinase D introduced the question of how much of the residual hexokinase D activity generally observed in diabetes or after fasting can be attributed to it. Moreover, were all of the high-affinity hexokinases really stable in these conditions? For the total liver that appeared to be true, but what about the 5% of high-affinity activity in the hepatocytes? All of these were pertinent questions in relation to the metabolic role of hexokinase D.

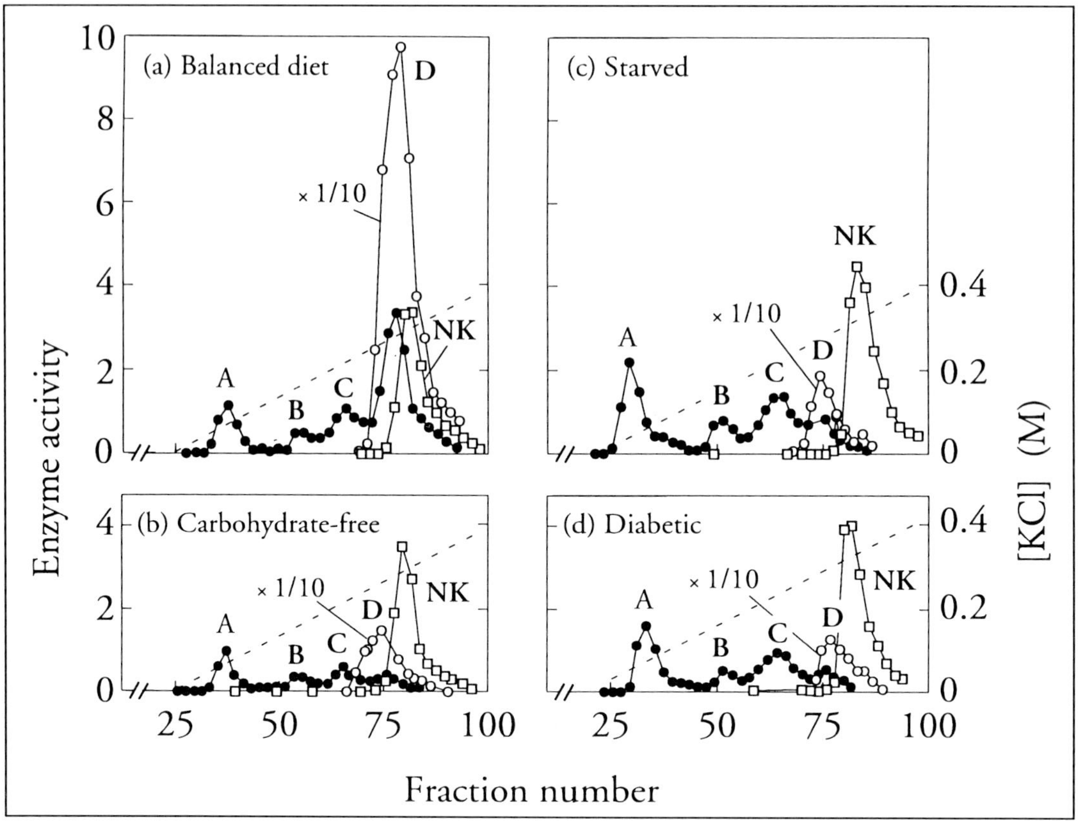

Fig. 4.3. Effect of diet on hexokinase isoenzymes in hepatocytes. The high level of hexokinase D (○, curves D) in animals fed a balanced diet (a) is greatly decreased in animals receiving a carbohydrate-free diet (b), or starved for 4 days (c), or after 4 days of streptozotocin-induced diabetes (d). The activities of the other hexokinase isoenzymes (●, curves A, B and C) and of N*-acetylglucosamine kinase (□, curves NK) were virtually constant. Other details were as for Figure 4.2. Reproduced with permission from Arch Biol Med Exper 1985; 18:331-358.*

When we investigated this problem we found that only the specific activity of hexokinase D showed changes, but not those of the three high-affinity isoenzymes, whether in hepatocytes or in nonparenchymal cells, under all conditions used.[30,31] Figure 4.3 shows the results obtained in hepatocytes from normal, diabetic and fasted rats, and from rats fed on a carbohydrate-free diet. A point to emphasize is that although the activity of hexokinase D in the liver is considerably decreased in all of these conditions, as expected, it does not disappear completely but reaches a final residual value of about 15-20% of normal,[30,31] indicating the establishment of a new steady state. Extending the period of a carbohydrate-free diet to 12 days produced no further decrease in hexokinase D activity, supporting this idea.

N-Acetylglucosamine kinase, on the other hand, showed no change in either hepatocytes or nonparenchymal cells under all conditions. However, it can be seen from Figure 4.3 that the residual hexokinase D activity measured at 100 mM glucose is clearly distinct from the phosphorylating activity attributable to *N*-acetylglucosamine kinase. Furthermore, this residual hexokinase D activity is still about 2-4-fold higher at physiological glucose concentrations in the range 5-15 mM than the sum of hexokinases A, B and C. Thus the residual hexokinase D activity and the high-affinity isoenzymes could both contribute to the glucose phosphorylation observed in hepatocytes isolated from diabetic adult rats.[19] It might be argued, however, that because of the concentration of glucose 6-phosphate present in the hepatocytes the high-affinity hexokinases would be inhibited, as appears to be the case in the pancreas (see Chapter 5). Nevertheless, the situation is not so clear, because of the apparently nonhomogeneous distribution of the low-K_m hexokinases: hexokinase C appears to be a perinuclear enzyme,[32] whereas hexokinase A is bound to the mitochondrial and microsomal membrane; moreover, there is evidence that glucose 6-phosphate is itself compartmentalized within the hepatocyte.[33] Thus one cannot assume automatically that all high-affinity hexokinases will detect an inhibitory concentration of glucose 6-phosphate.

The residual hexokinase D activity is in accord with the residual concentration of translatable mRNA coding for rat liver hexokinase D in starved and diabetic animals.[34] This is further discussed in Chapter 8.

Does hexokinase D coexist with glucose-6-phosphatase in the hepatocyte?

An important physiological problem is how much substrate cycling between glucose and glucose 6-phosphate occurs in liver, in particular in the hepatocytes, where hexokinase D is located. To maintain glucose homeostasis the liver removes glucose during the absorptive phase after a normal meal for the synthesis of glycogen and triglycerides, and it releases glucose during the postabsorptive phase, mainly for use by the central nervous system and erythrocytes. Thus the parenchyma catalyzes glycogen synthesis, glycolysis and liponeogenesis during the absorptive phase, but gluconeogenesis during the postabsorptive phase. (Glycolysis is part of the conversion of glucose to fatty acids and thus is functionally linked to liponeogenesis).

In the postabsorptive and fasting states, therefore, glycogenolysis and gluconeogenesis provide a steady-state supply of glucose 6-phosphate, which is converted into free glucose by glucose 6-phosphatase for release into the circulation. As hexokinase D and glucose 6-phosphatase catalyze opposing reactions, the net carbon flux between glucose and glucose 6-phosphate, equivalent to the net movement of glucose into or out of the liver, is determined by the differences in the rates of the two reactions. These depend on the amounts of the two enzymes and on the balance of all the different factors that may affect them. In the liver as a whole there is no doubt that a substrate cycle exists, at least in the rat, as we saw above that hexokinase D does not disappear completely in fasting or diabetes. The degree of cycling that occurs in hepatocytes may nonetheless be minor as a consequence of metabolic zonation, or functional specialization of individual hepatocytes.

Metabolic zonation in liver

The model of "metabolic zonation" proposes that the two antagonistic pathways are spacially separated in periportal and perivenous hepatocytes,[35,36] in the same way that glycolysis and gluconeogenesis are separated in distal and proximal kidney tubules. Two different zones can be recognized in liver according to the flow of blood: the *periportal zone* (also known as the *afferent* or *proximal* zone), perfused with blood rich in oxygen, substrates and hormones; and the *perivenous zone* (*pericentral*, *centrilobular*, *efferent* or *distal* zone), perfused with blood depleted in oxygen, substrates and hormones, but enriched in CO_2 and other products of metabolism. (The autonomic innervation of the two zones may also be different.) The upstream and downstream zones differ in that they contain hepatocytes with different contents of enzymes and subcellular structures, and they contain different numbers of non-parenchymal cells, which are predominantly localized in the periportal zone.[36]

The concept of metabolic zonation is supported by the observation that the gluconeogenic enzymes phosphoenolpyruvate carboxykinase, fructose-bisphosphatase and glucose-6-phosphatase are predominantly localized in the upstream, periportal domain of the liver lobule, whereas the glycolytic enzyme pyruvate kinase and the glucose phosphorylating activity are in the downstream, perivenous zone of the liver acinus (see the review of Jungermann and Katz for specific references).[36] Concentration gradients for oxygen, substrates and hormones in the liver microcirculation are thought to be responsible for region-specific gene expression, resulting in a topological separation of metabolic pathways. The borders of the two domains are not fixed as they depend on the metabolic state of the animal.

In the case of hexokinase D, there is immunohistochemical and microdissection evidence that its concentration is approximately 2-3 times higher in perivenous than in periportal liver tissue,[37-42] whereas glucose 6-phosphatase varies in the opposite direction.[43-45] The localization of hexokinase D is intriguing, because if its main function is the uptake of the glucose that enters the liver during the absorptive phase one would expect it to predominate in the periportal region and not in the perivenous. Its metabolic role may perhaps be more complex than has yet been realized.

The localization of hexokinase D mRNA has been studied by in situ hybridization under normal and inducing conditions. In contrast to the predominant localization of the enzyme in the perivenous region of adult liver rat, its mRNA seems to be homogeneously distributed in the liver lobule,[46] as with the mRNA of pyruvate kinase L. However, in the inducing conditions that exist during postnatal development, after an oral glucose load or dexamethasone treatment, there is definitely a transient predominantly pericentral pattern of expression of hexokinase D mRNA, similar to the distribution of the protein, suggesting a pretranslational level of regulation in the perivenous area.[46] The molecular mechanisms underlying the zonation of gene expression are only partly understood, but hybridocytochemical analyses have disclosed that the major site of control is at the pretranslational level.

The enzyme distribution suggests that in normal rat liver glucose release predominates in the periportal domain and glucose uptake in the pericentral domain. This implies that the periportal hepatocytes release glucose formed from pyruvate via gluconeogenesis as well as from glycogen synthesized mainly from three-carbon substrates (indirect pathway), whereas the perivenous hepatocytes use glucose for the synthesis of glycogen (direct pathway) and triglycerides via glycolysis and lipogenesis, as illustrated in Figure 4.4.

There have been attempts to estimate the zonal flux rates from the zonal concentrations of glucose and glucose 6-phosphate and the zonal activities and half-saturation concentrations of the two enzymes hexokinase D and glucose 6-phosphatase, assuming that the cycle would be controlled only by substrate availability. Such calculations have shown that during a normal 24-hour feeding cycle the periportal zone should always release glucose, although in variable amounts according to the conditions; conversely, the perivenous zone should always take up glucose. The expected shift from net glucose uptake to net glucose release when the animals progress from feeding to fasting would result from the average metabolic activities in the periportal and the perivenous zone.[47] Of course, the results of such a calculation can only be approximate as neither the cooperativity of hexokinase D nor its possible allosteric control were taken into account (see Chapter 7), nor any possibility of control of glucose 6-phosphatase. Another complication is the possible compartmentation of glucose 6-phosphate.[33]

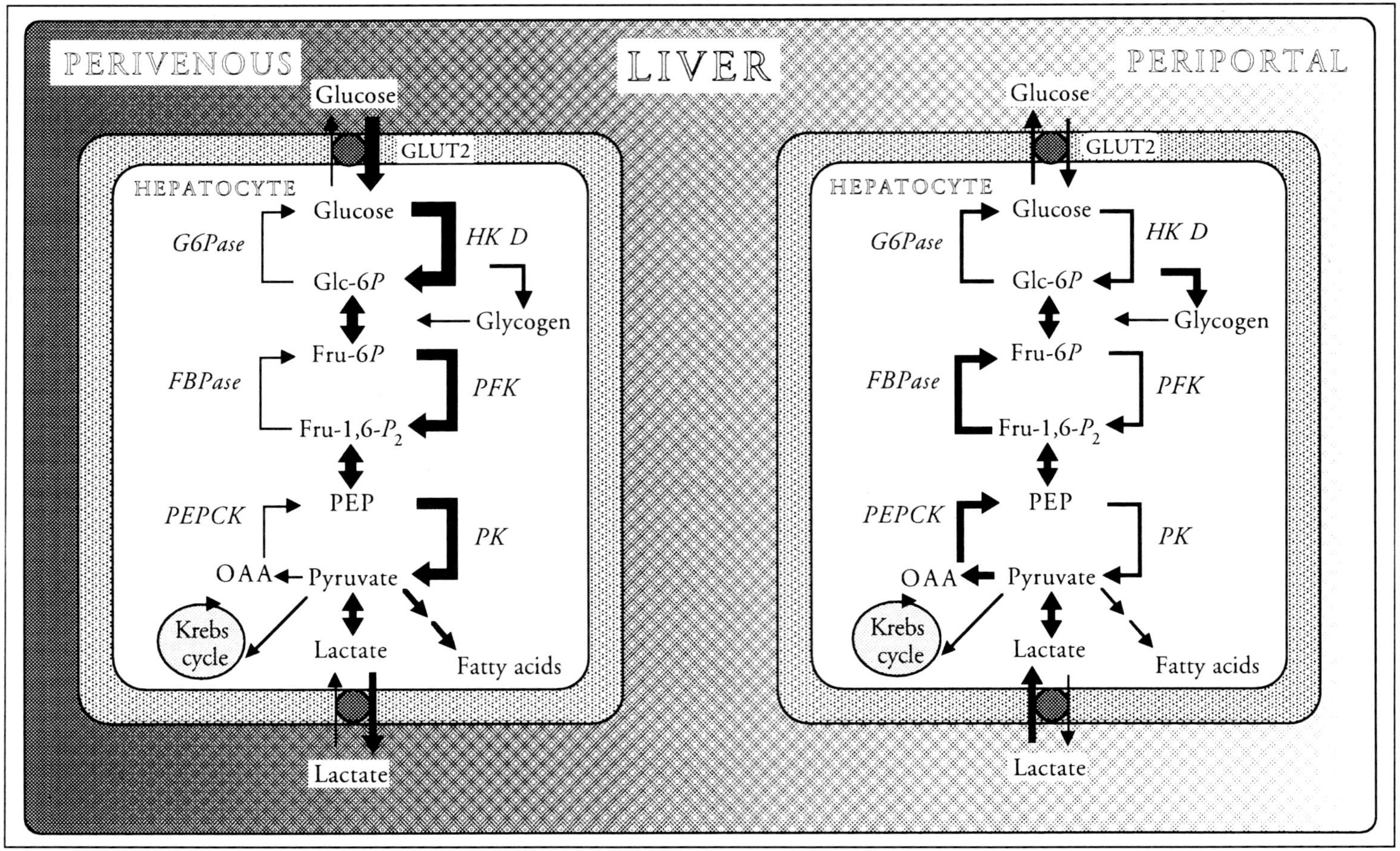

Fig. 4.4. *Zonation of hepatocytes. Hepatocytes in the perivenous domain of the liver have high activity for glucose uptake and glycolysis, whereas those in the periportal domain have these activities reduced in favor of higher activity in gluconeogenesis. As indicated by the continuous gradient of the background shading, the borders between the two domains are indefinite and vary with the metabolic state. Redrawn from Biochem J 1993; 293:1-13.*

PERSPECTIVES

In this chapter we have seen that hexokinase D in liver is confined to hepatocytes. It predominates in the perivenous zone, where there appears to be a net uptake of glucose, which is converted to glycogen by the direct pathway, as well as a net synthesis of fatty acids from glucose, which does not occur in the periportal zone. Hexokinase D has been classically considered as a soluble enzyme, not bound to membranes, and our studies in isolated hepatocytes, mentioned earlier in this chapter, support this view. (For a while, I was quite attracted by the idea of a possible interaction with glycogen particles, but there is still no evidence for this.) However, recent studies have indicated that the situation may be more complex, as rat hepatocytes cultured in 5 mM glucose show hexokinase D activity to be present predominantly in a bound state. Thus less than 20% of hexokinase D activity is released by permeabilization of the cells with digitonin in the presence of Mg^{2+}.[48]

The proportion of bound activity is affected by treatment of hepatocytes with cytochalasin-D or phalloidin, which alter the conformation of microfilament cytoskeleton, and on this basis it was postulated that hexokinase D may occur in association with the microfilament cytoskeleton. The enzyme can be released or translocated from its bound state by incubation of hepatocytes with higher glucose concentration (15-35 mM) or with fructose at very low concentrations (50μM). This translocation is inhibited by cyanide, suggesting the participation of an ATP-dependent mechanism.[48] The effects of glucose and fructose were rapid (with a half-time of 5 min) and reversible, and were potentiated by insulin and counteracted by glucagon. Sorbitol is also very effective, being six-fold more potent than fructose.[49]

A concentration of 15 mM glucose causes half-maximal activation whereas a concentration of fructose as small as 50 μM produces complete activation. The fact that the required concentration of fructose is so much lower than that of glucose suggests that a metabolite generated via fructokinase could be involved. Although Agius also suggests other possibilities,[48] it is tempting to speculate that the bound state of hexokinase D could be mediated by the regulatory protein, which will be treated in Chapter 7, as the concentration of fructose that activates glucose phosphorylation in intact hepatocytes is the same as the concentration that stimulates translocation. Nevertheless, the bound state of hexokinase D is a recent discovery, and the mechanism and its control are far from clear. One thing that appears to be clear, however, is that the stimulation of hexokinase D translocation is associated with a corresponding stimulation of conversion of glucose into glycogen.

It will therefore be very useful to have new estimates of the zonal flux rates in the cycle between glucose and glucose 6-phosphate that take account of the glucose cooperativity, the allosteric inhibition by acyl-CoA and the regulatory protein, and the apparent necessity of hexokinase D to be translocated in order to be active.

References

1. Colowick SP. The hexokinases. In: Boyer PD, ed. The Enzymes. 3rd ed. New York: Academic Press, 1973; 9:1-48.
2. Davagnino J, Ureta T. The identification of extrahepatic "glucokinase" as *N*-acetyl-glucosamine kinase. J Biol Chem 1980; 255:2633-2636.
3. Allen MB, Brockelbank JL, Walker, DG. Apparent glucokinase activity in non-hepatic tissues due to *N*-acetyl-D-glucosamine kinase. Biochim Biophys Acta 1980; 614:357-366.
4. Vera ML, Cárdenas ML, Niemeyer H. Kinetic, chromatographic and electrophoretic studies on glucose-phosphorylating enzymes of rat intestinal mucosa. Arch Biochem Biophys 1984; 229:237-245.
5. Ashcroft SJ, Randle PJ. Enzymes of glucose metabolism in normal mouse pancreatic islets. Biochem J 1970; 119:5-12.
6. Burch PT, Trus M, Berner DK et al. Adaptation of glycolytic enzymes. Glucose use and insulin release in rat pancreatic islets during fasting and refeeding. Diabetes 1981; 30:923-928.
7. Trus MD, Zawalich KC, Burch PT et al. Regulation of glucose metabolism in pancreatic islets. Diabetes 1981; 30:911-922.
8. Meglasson MD, Burch PT, Berner DK et al. Chromatographic resolution and kinetic characterization of glucokinase from islets of Langerhans. Proc Natl Acad Sci USA 1983; 80:85-89.
9. Meglasson MD, Burch PT, Hoenig M et al. Identification and significance of glucokinase in transplantable insulinomas. J Biol Chem 1983; 258:2094-2097.
10. Iynedjian PB, Möbius G, Seitz HJ et al. Tissue-specific expression of glucokinase: identification of the gene product in liver and pancreatic islets. Proc Natl Acad Sci USA 1986; 83:1998-2001.
11. Iynedjian PB, Pilot P-R, Nouspikel T et al. Differential expression and regulation of the glucokinase gene in liver and islets of Langerhans. Proc Natl Acad Sci USA 1989; 86:7838-7842.
12. Hughes SD, Quaade C, Milburn JL et al. Expression of normal and novel glucokinase mRNAs in anterior pituitary and islet cells. J Biol Chem 1991; 266:4521-4530.
13. Liang Y, Jetton TL, Zimmerman EC et al. Effects of alternate RNA splicing on glucokinase isoform activities in the pancreatic islet, liver and pituitary. J Biol Chem 1991; 266:6999-7007.
14. Jetton TL, Liang Y, Pettepher CC et al. Analysis of upstream glucokinase promoter activity in transgenic mice and identification of glucokinase in rare neuroendocrine cells in the brain and gut. J Biol Chem 1994; 269:3641-3654.
15. Gould GW, Bell GI. Facilitative glucose transporters: an expanding family. Trends Biochem Sci 1990; 15:18-23.
16. Greengard O, Federman M, Knox WE. Cytomorphometry of developing rat liver and its application to enzymic differentiation. J Cell Biol 1972; 52:261-272.
17. Sapag-Hagar M, Marco R, Sols A. Distribution of hexokinase and glucokinase between parenchymal and non-parenchymal cells of rat liver. FEBS Lett 1969; 3:68-71.

18. Crisp DM, Pogson CI. Glycolytic and gluconeogenic activities in parenchymal and non-parenchymal cells from mouse liver. Biochem J 1972; 126:1009-1023.
19. Bontemps F, Hue L, Hers H-G. Phosphorylation of glucose in isolated rat hepatocytes. Sigmoidal kinetics explained by the activity of glucokinase alone. Biochem J 1978; 174:603-611.
20. Morrison GR. Hexokinase and glucokinase activities in bile duct epithelial cells from normal rat and human livers. Arch Biochem Biophys 1967; 122:569-573.
21. Werner HV, Bartley JC, Berry MN. Glucose-adenosine 5'-triphosphate 6-phosphotransferases of isolated rat liver parenchymal cells. Biochem J 1972; 130:1153-1155.
22. Bonney RJ, Walker PR, Potter V. Isoenzyme patterns in parenchymal and non-parenchymal cells isolated from regenerating and regenerated rat liver. Biochem J 1973; 136:947-954.
23. Seglen PO. Preparation of isolated liver cells. In: Prescott M, ed. Methods in Cell Biology, New York: Academic Press, 1976; 13:28-83.
24. Reyes A, Cárdenas ML. All hexokinase isoenzymes coexist in heptocytes. Biochem J 1984; 221:303-309.
25. Radojkovic J, Slebe JC, Ureta T. A radioassay for the glucokinase reaction. Arch Biol Med Exper 1978; 11:73-76.
26. Viñuela E, Salas M, Sols A. Glucokinase and hexokinase in liver in relation to glycogen synthesis. J Biol Chem 1963; 238:PC1175-1177.
27. Machiya T, Hoyosa N. Reaction mechanism of hexokinase isozymes and their insulin response. Endocrinol Jpn 1969; 16:433-445.
28. Pilkis SJ. Hormonal control of hexokinase activity in animal tissues. Biochim Biophys Acta 1970; 215:461-476.
29. Wakelam MJ, Aragón C, Giménez C et al. Thyroid hormones and the precocious induction of hepatic glucokinase in the neonatal rat. Eur J Biochem 1979; 100:467-475.
30. Reyes A, Rabajille E, Cárdenas ML et al. Stability of hexokinases A, B, and C, and *N*-acetylglucosamine kinase in liver cells isolated from rats submitted to diabetes and several dietary conditions. Biochem J 1984; 221:311-315.
31. Niemeyer H, Cárdenas ML. Concurrencia de mecanismos múltiples e integrados en la modulación de la actividad enzimática: significado para la regulación de flujos metabólicos. Arch Biol Med Exper 1985; 18:331-358.
32. Preller A, Wilson JE. Localization of the type III isozyme of hexokinase at the nuclear periphery. Arch Biochem Biophys 1992; 294:482-492.
33. Christ B, Jungermann K. Sub-compartmentation of the "cytosolic" glucose-6-phosphate pool in cultured rat hepatocytes. FEBS Lett 1987; 221:375-380.
34. Spence JT. Levels of translatable mRNA coding for rat liver glucokinase. J Biol Chem 1983; 258:9143-9146.
35. Jungermann K, Sasse D. Heterogeneity of liver parenchymal cell cells. Trends Biochem Sci 1978; 3:198-202.
36. Jungermann K, Katz N. Functional specialization of different hepatocyte populations. Physiol Rev 1989; 69:708-764.

37. Katz N, Teutsch HF, Jungermann K et al. Heterogeneous reciprocal localization of fructose-1,6-bisphosphatase and glucokinase in microdissected periportal and peri-venous rat liver tissue. FEBS Lett 1977; 83:272-276.
38. Trus M, Zawalisch H, Gaynor D et al. Hexokinase and glucokinase distribution in the liver lobule. J. Histochem. Cytochem 1980; 28:579-581.
39. Fischer W, Ick M, Katz N. reciprocal distribution of hexokinase and glucokinase in periportal and perivenous rat liver tissue. Hoppe-Seyler's Z Physiol Chem 1982; 363:375-382.
40. Teutsch HF, Lowry OH. Sex-specific differences in hepatic glucokinase activity. Biochem Biophys Res Comm 1982; 106:533-538.
41. Lawrence GM, Jepson MA, Trayer IP et al. The compartmentation of glycolytic and gluconeogenic enzymes in rat kidney and liver and its significance to renal and hepatic metabolism. Histochem J 1986; 18:45-53.
42. Lawrence GM, Trayer IP, Walker DG. Histochemical and immunohistochemical localization of hexokinase isoenzymes in normal rat liver. Histochem J 1984; 16:1099-1111.
43. Katz N, Teutsch HF, Sasse D et al. Heterogeneous distribution of glucose-6-phosphatase in microdissected periportal and perivenous rat liver tissue. FEBS Lett 1977; 76:226-230.
44. Teutsch HF. Quantitative determination of G6P'ase activity in histochemically defined zones of the liver acinus. Histochemistry 1978; 58:281-288.
45. Teutsch HF. Chemomorphology of liver parenchyma. Qualitative histochemical distribution patterns and quantitative sinusoidal profiles of G6Pase, G6PDH and malic enzyme activity and of glycogen content. Prog Histochem Cytochem 1981; 14:1-92.
46. Moorman AFM, de Boer PAJ, Charles R et al. Pericentral expression pattern of glucokinase mRNA in the rat liver lobulus. FEBS Lett 1991; 287:47-52.
47. Jungermann K, Heilbronn R, Katz N et al. The glucose/glucose 6-phosphate cycle in the periportal and perivenous zones of rat liver. Eur J Biochem 1982; 123:419-436.
48. Agius L, Peak M. Intracellular binding of glucokinase in hepatocytes and translocation by glucose, fructose and insulin. Biochem J 1993; 296:785796.
49. Agius L. Control of glucokinase translocation in rat hepatocytes by sorbitol and the cytosolic redox state. Biochem J 1994; 298:237-243.

CHAPTER 5

Pancreatic and Liver Hexokinase D: One Gene Two Proteins

Hexokinase D is the characteristic hexokinase of the hepatocytes, but it also exists in pancreatic β-cells, where it has a very important metabolic role to play. Although this book deals mainly with the hepatic form of the enzyme, recent years have seen some very important discoveries concerning the pancreatic form. Although the hexokinases D from pancreas and liver are similar in kinetic behavior and are coded by the same gene, their primary structures in the *N*-terminal regions are different because of different splicing of the RNA transcript. The use of distinct tissue-specific promoters and alternative first exons in the gene results in hexokinase D mRNA in the liver and islet that differ at their 5' ends. Another very important discovery, which has greatly stimulated research on hexokinase D, is the association between the mutations in the hexokinase D gene and an autosomal dominant form of non-insulin-dependent diabetes mellitus (NIDDM), confirming an earlier hypothesis of Matschinsky.[1] This chapter is a brief account of the discovery of the pancreatic enzyme, its physiological role, the structure of the hexokinase D gene and the linkage between hexokinase D and NIDDM. Pertinent reviews are those of Iynedjian,[2] Permutt and coworkers,[3] Matschinsky,[4] Randle,[5] and Pilkis and coworkers.[6]

PANCREATIC HEXOKINASE D

Discovery: *N*-acetylglucosamine kinase or hexokinase D?

As we have seen in the previous chapter, the two decades after the first studies saw several reports of a hexokinase with low sensitivity to glucose in tissues apart from liver, suggesting the presence of hexokinase D in those tissues; however, the clear demonstration that this activity was due to the presence of *N*-acetylglucosamine kinase and not of hexokinase D left the feeling that hexokinase D was confined to hepatocytes. A doubt persisted, however, in relation to the pancreas, and in particular to the β-cells, the cells that produce insulin: did they contain hexokinase D? The islets of Langerhans, the endocrine part of pancreas, contain phenotypically distinct cells that produce insulin (β-cells), glucagon (α-cells), somatostatin (D-cells) and pancreatic polypeptide (PP cells). The presence of a phosphorylating system of high K_m for glucose had been reported in islets, and the question was whether this also was a *N*-acetylglucosamine kinase or a hexokinase D. This had not been resolved at the time of the studies discussed in Chapter 4.

On one hand, various groups had reported hexokinase D in pancreatic β-cells in islets of, for example, the rabbit,[7] the obese hyperglycemic mouse,[8] the normal mouse,[9,10] and the rat.[11] On the other hand, many investigators had been unable to detect hexokinase D in islets,[12] and the reported V_{max} and K_m values both varied over ten-fold ranges, reminiscent of the artifactual reports of hexokinase D in other tissues due to the presence of *N*-acetylglucosamine kinase, as already discussed. Moreover, the likelihood of misidentification increased when *N*-acetylglucosamine kinase was found to be present in islets.[13]

Establishing the truth about the presence of hexokinase D in islets was important for understanding the physiology of the pancreas, as Matschinsky had postulated that hexokinase D could be a glucose sensor (an adaptive glucoreceptor), relating ambient glucose to insulin secretion.[8] (A more recent and explicit expression of the idea that hexokinase D is the component allowing the β-cell to recognize and measure the fluctuations of blood glucose may be found in Matschinsky's review.)[4] It may seem curious, therefore, that the identities of the glucose-phosphorylating enzymes in pancreas were not determined at the same time as the various false trails in other less pertinent tissues were explored. However, the laborious procedures necessary for isolating pancreatic islets from rats and the limited yields from very large numbers of animals meant that the work had to be left to a group with the necessary experience in working with the pancreas; it was not by chance, therefore, that it was Matchinsky's group in Philadelphia that solved the problem.

They were interested in the general problem of diabetes and insulin production, and they reinvestigated the presence of hexokinase D in pancreas and the control of glycolytic flux.[14,15] Finally, they were able to show that β-cells[16] and insulinomas[17] contained hexokinase of low affinity for glucose that was not *N*-acetylglucosamine kinase, but the exocrine pancreas did not. Much later Iynedjian and colleagues used immunoblotting to show the presence of both hexokinase D and its mRNA in pancreatic islets;[18,19] more recently immunocytochemical studies have allowed them to be localized in islet β-cells.[20]

Properties of pancreatic hexokinase D

Pancreatic hexokinase D was found to have the same kinetic characteristics as liver hexokinase D, namely kinetic cooperativity with respect to glucose, with a Hill coefficient of about 1.6, and a $K_{0.5}$ of about 5 at pH 8.0.[16,17] Moreover, it could phosphorylate glucose, mannose, 2-deoxyglucose and fructose, but not *N*-acetylglucosamine, and the phosphorylation of mannose was also cooperative.[21,22] Its molecular mass, the amino acid composition and the chromatographic mobility were also similar to those of the liver enzyme. Finally, electrophoretic mobility in one- and two-dimensional gel systems and peptide maps after limited proteolysis resembled those for liver hexokinase D.[16,18,23] All of this suggested that pancreatic and liver hexokinase D were the same enzyme, coded by the same gene.

Despite all these similarities there was a striking difference between the liver and pancreatic enzymes. The great sensitivity of the activity of hepatic hexokinase D to dietary and hormone factors that we have discussed in Chapter 1 was not repeated for the pancreatic enzyme, which appeared to be constitutive, only a very minor variation in activity (about 30%) being observed after 72 h fasting.[15] This has been confirmed in more recent work.[19] Taken together, the results suggested that the liver and pancreatic enzymes were probably the product of the same gene, but that they were regulated differently. This was not in itself surprising, because, for example, hexokinase B is dependent on insulin in muscle but not in liver.

The discovery of the cooperative behavior of pancreatic hexokinase D was welcome, because it had been known for some time that glucose utilization by islets and insulin secretion by β-cells showed a sigmoidal dependence on the glucose concentration;[24-26] the half-maximum rate of secretion in the rat was obtained at 6-10 mM glucose,[27,28] and the half-maximum rate of glycolysis, for the low-affinity component of glucose usage, occurred at 11 mM glucose.[14] These and other facts[14,16,17] showed that pancreatic hexokinase D had properties making it a plausible candidate for the determining enzyme in glucose-stimulated insulin release. It is interesting to mention that the dose-response curve of glucose-induced proinsulin biosynthesis is also sigmoidal in purified rat β-cell preparations,[29] which agrees with the idea that glucose phosphorylation by hexokinase D would be the control point. (It has been claimed that this sigmoidicity could be a due to heterogeneity of the β-cell population,[30] but although the heterogeneity is probably real[20,30] it cannot generate sigmoidicity. On the contrary, any such heterogeneity must produce or increase an apparent *negative* cooperativity, as in the case of a heterogeneity of enzyme molecules or active sites discussed in Chapter 3.) It must be noted, however, that the concentration dependences of glucose usage and insulin secretion are not identical, the cooperativity in the stimulation of insulin secretion being greater than it is for the low-affinity component of glucose usage, which is similar to that of purified hexokinase D (Hill coefficient of 1.5-1.6 in both cases). Thus the cooperativity of hexokinase D only partially explains the cooperativity of insulin secretion, and part of it could be due to later steps; given the complexity of the whole process (see Fig. 5.1),[30] this is hardly surprising.

Physiological role of pancreatic hexokinase D

The β-cells of the islets of Langerhans act as fuel homeostats in mammals. They detect changes in the availability of nutrient molecules such as glucose, amino acids, fatty acids and ketone bodies, and respond by secreting insulin at an appropriate rate. At the concentrations of the nutrients that occur in plasma only glucose is capable of initiating insulin secretion.[28] Under physiological conditions glucose both stimulates insulin secretion and permits potentiation of insulin secretion by the other nutrients mentioned, and the degree of potentiation possible increases with the glucose concentration.

Thus the detection of hexokinase D in the β-cells producing insulin was metabolically very important because it suggested a link between glucose phosphorylation by hexokinase D and insulin production. As a result, as mentioned above, Matchinsky and his colleagues introduced the idea of hexokinase D as a glucose sensor, a concept that has accumulated support through the years.

The single most important stimulus for insulin secretion by the β-cells appears to be a rise in the concentration of glucose in the blood, which must be metabolized; thus, glucose-induced insulin secretion is closely associated with an increase of glycolytic flux in islets.[25] The requirements for the insulinotropic effect include the cellular uptake of glucose, and its metabolism via glycolysis and the Krebs cycle.[13] Consequently, the β-cells can respond adequately to physiological variations of glucose and accordingly varying their secretion of insulin only if certain conditions are met: (i) it should be possible to establish a rapid equilibrium between the glucose concentration in the extra- and intra-cellular compartments; (ii) the enzyme that catalyzes the first step in glucose metabolism ought to be able to vary its activity in response to physiological variations in the glucose concentrations, and ought to control the flux through the metabolic pathway so that the variations in its activity could be transmitted to the rest of the pathway; and (iii) a link must exist between the rate of glucose metabolism and the insulin release process.

(i) Equilibrium between extra- and intra-islet glucose concentration

Glucose apparently does equilibrate rapidly across the cell membrane,[31] and so the first condition appears to be satisfied. In support of these early observations the glucose transporter isotype normally expressed in islet β-cells is GLUT2.[32-34] This is a very efficient transporter, present also in liver, and the maximal rates of glucose transport measured in isolated islets or dispersed β-cells exceed the rates of glucose metabolism approximately 100-fold, indicating that metabolism is not controlled at the plasma membrane transport step.[31,35] Furthermore, its high half-saturation concentration, about 17 mM, presumably facilitates rapid reequilibration of the intra-islet concentration when the blood glucose concentration is altered.[33] Efrat and coworkers give further discussion of the transporter and the relationship between it and hexokinase D as glucose sensors.[36]

(ii) Sensitivity to variations in glucose concentration and control of the metabolic flux

The first step in glucose metabolism is phosphorylation. Apart from hexokinase D β-cells have a phosphorylating system of low K_m,[37] indicating the presence of one or more hexokinases of high affinity for glucose, but these would be saturated at the physiological glucose concentrations and consequently would not be good candidates. In islet homogenates Matschinsky's group[14] estimated a K_m of 0.03 mM glucose for the supernatant and 0.05 mM for the pellet, and Malaisse's group obtained a similar value.[37] Hexokinase D, on the other hand, has kinetic properties that would allow it respond sensitively to physiological variations of the glucose concentration. (See Chapter 6 for further discussion.) So the question is how much of the glucose that arrives in β-cells is phosphorylated by hexokinase D and how much by the other hexokinases.

This was a difficult question, as there appeared to be a discrepancy between the enzymic and metabolic data; experiments with islet homogenates at physiological glucose concentrations had shown that glucose phosphorylation was mainly (about 80%) attributable to low-K_m hexokinases;[8-11] in intact islets, however, glucose utilization appeared to depend on a system of high K_m. Further study confirmed that the system of high K_m accounted for only 25% to 30% of the total hexokinase activity in islet extracts, but that it appeared to be the major contributor, approximately 85%, to the total glucose usage in intact islet cells.[14] Thus glucose use occurs with maximum rate and half-maximum glucose concentration resembling those of hexokinase D, suggesting that the hexokinase of low K_m was inhibited. The authors interpreted this result in terms of the effective inhibition of the low-K_m hexokinases by the ambient glucose 6-phosphate concentration. Later on, with the idea of testing this hypothesis to explain the disparity between the metabolic and enzymic results, Malaisse's group measured the steady-state content of glucose 6-phosphate at increasing concentrations of extracellular glucose, up to 28 mM, finding values of 0.26 to 0.79 pmol/islet.[37] From these measurements and an estimated intracellular space in the islets of 1.19 nl/islet, one can estimate a cellular concentration of glucose 6-phosphate varying between approximately 0.2 and 0.7 mM, similar to the concentrations reported in liver. If there is compartmentation the local concentration could be higher.

As the constant for inhibition of glucose phosphorylation by glucose 6-phosphate in islet homogenates was close to 0.13 mM,[37] it is sensible to argue that the high-affinity hexokinase activity would be inhibited at normal physiological glucose concentration. Moreover, the high-affinity hexokinase activity in islet homogenates is also inhibited by glucose 1,6-bisphosphate with a K_i value of 0.2 mM.[37] However, although this interpretation appears quite reasonable one has to bear in mind that if the phosphorylation step controls the pathway the level of the intermediate glucose 6-phosphate must be very low (the bigger the control exerted by the phosphorylation step the smaller ought to be the concentration of glucose 6-phosphate), though this could, of course, be counteracted by a gluconeogenic pathway. Several lines of evidence seem to suggest that the phosphorylation of glucose by hexokinase D is the step that controls the glycolytic flux during hyperglycemia in whole islets and by implication in β-cells, which constitute approximately 80% of the islet mass.[38] Glucose use by intact islets has a maximal rate that is closer to the maximal rate of hexokinase D than to the maximal rate of other enzymes in the pathway. Enzyme assays under limiting rate conditions indicate that total hexokinase activity, including both low-K_m hexokinases and hexokinase D, is much lower than the activities of the other two enzymes that are good candidates to regulate the pathway, i.e. 6-phosphofructo-1-kinase and pyruvate kinase.[14] Although this could be an argument in favor of the idea that the control resides in the phosphorylation step, one has to be cautious; to establish how much control an enzyme exerts on a pathway it is not enough to measure maximal activities of each step in isolation. One needs to know the activity in the presence of all the metabolites.

All the data taken together suggest that hexokinase D has control over the carbon flux in the glycolytic pathway, allowing physiological fluctuations of glucose concentration to translate into alterations of the overall glycolytic rate. (See the article by Efrat and coworkers[36] for further discussion).

(iii) Link between glucose metabolism and insulin secretion

The coupling link between glycolytic metabolism and insulin secretory activity is thought to reside in ATP-sensitive potassium channels (K^+ATP) in the plasma membrane of the β-cell.[39,40] It has been proposed that a localized increase in the intracellular ATP/ADP ratio leads to closure of these channels. The ensuing membrane depolarization and electrical activity cause the opening of voltage-dependent calcium channels. The influx of extracellular calcium ions into the cell sets in motion the events leading to the exocytotic release of insulin from secretory granules by steps that remain to be elucidated in detail. These are summarized in Figure 5.1, which is based on information in Figure 5 of reference 2 and Figure 1 of reference 30. The increased intracellular free Ca^{2+} concentration may also act on mitochondrial dehydrogenases and respiratory chain phosphorylation to sustain the elevated ATP/ADP ratio.[41]

In summary, the responsiveness of the entire system to glucose would be based on hexokinase D, in agreement with the notion of a

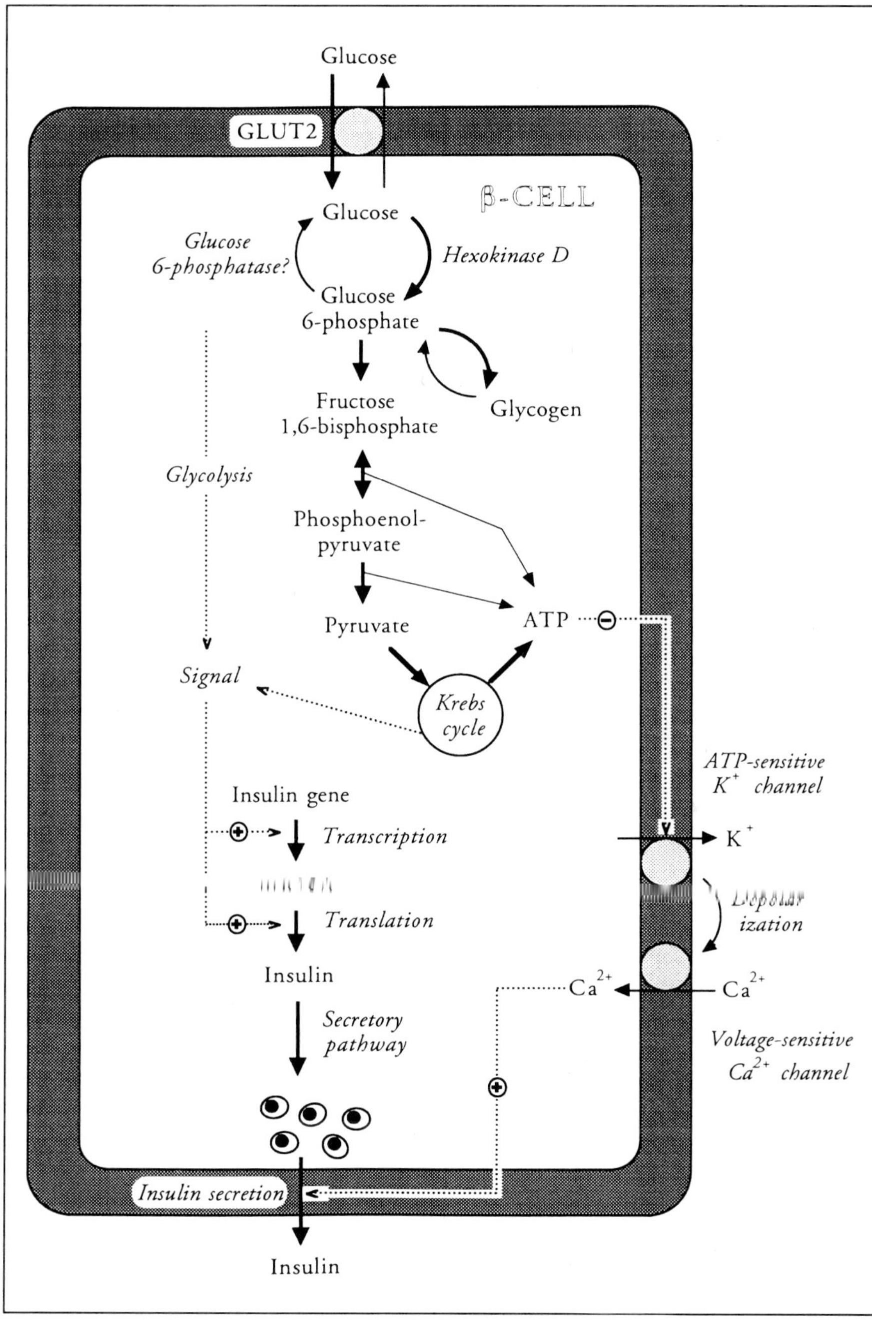

Fig. 5.1. Role of hexokinase D in β-cells of the islets of Langerhans. The scheme represents the complex series of signals that connect glucose uptake into the β-cell with secretion of insulin. A large amount of data[2,30] suggest that the hexokinase D step accounts for the major part of the control of metabolic flux, and that the increased production of ATP that follows increased glucose uptake causes membrane depolarization and opening of voltage-sensitive calcium channels and an influx of Ca^{2+} ions necessary for insulin release; at the same time metabolic signals are converted into genetic signals causing increased insulin synthesis.

glucose sensor used by Matschinsky to refer to pancreatic hexokinase D. However, very recent experiments with transgenic mice,[42] described at the end of this chapter, appear to indicate that the level of hexokinase D could be decreased in pancreas to 30% without affecting the level of insulin and glucose in the plasma, which indicates that the problem in the whole animal could be quite complex. This is not surprising if one recalls that glucose-induced insulin secretion is also subject to stimulatory and inhibitory modulation by hormones and neurotransmitters.[30]

It is possible that metabolic zonation similar to that found in the liver (Chapter 4) may also occur in islets. For example, a recent immunohistochemical study has suggested large variations in the amounts of hexokinase D among individual β-cells of the same islet.[20]

Activity of glucose-6-phosphatase and a immunoreactive protein has been detected in extracts of whole islets.[43] However, activity of glucose 6-phosphatase in intact islets may be less than activities measured in islet homogenates, as this enzyme is inhibited by glucose at physiological concentrations. Its presence agrees with the substrate cycle observed between glucose and glucose 6-phosphate,[44,45] but the apparent zonation raises the question of whether hexokinase D and glucose-6-phosphatase coexist in the same cells and whether glucose cycling plays a role in the regulation of insulin secretion. Moreover, although glucose 6-phosphatase is present in islets of some rodent species, it is not present in all.[46]

An interesting observation in relation to glucose phosphorylation and insulin secretion is the one made by Iynedjian's group, who studied an insulinoma cell line (RINm5F) that produces insulin but shows no glucose-stimulated insulin release within the physiological range. This cell line has near-normal levels of hexokinase D, but dramatically overexpresses hexokinase B. The authors have proposed that the imbalance between hexokinase D and the low-K_m hexokinase activity is responsible for this impairment.[23]

ONE GENE, TWO PROTEINS

The development of various techniques of molecular biology, such as gene cloning and nucleic acid sequencing, made it possible to answer some of the remaining questions and to confirm the early studies. Hexokinase D cDNAs were first cloned from rat liver,[47] and shortly afterwards from both pancreas and pituitary gland.[48-50] Southern blots of rat genomic DNA using a cDNA probe of hepatic hexokinase D showed only a single copy of the gene,[47] confirming what had been suspected. Similarly only one hexokinase D gene is detected in humans, in band p13 of chromosome 7.[51] Cloning of the cDNA for hexokinase D from liver and pancreatic islets allowed the amino acid sequences to be deduced, and these are now known for the enzyme from both liver and pancreatic islet β-cells in both rat and human (Fig. 5.2).[47,50-55] Comparison of cDNA between liver and islet showed

N-terminal sequence differences in 15 amino acids, which was unexpected in view of the existence of just one gene (see below). Liver and islet enzymes from human and rat contain 465 amino acids,[47,54] with *N*-termini differing by 13-15 residues. Determination of the gene structures for the two rat enzymes showed that both cDNAs are coded by a single gene with 10 exons, but that, as we shall see shortly, exon 1

CCTTGGCTGCCTATCTTTTGCAACACTCAGCCAGACAGTCC

TCACCTGCAACAGGTGGCCTCAGGAGTCAGGAACATCTCTACTTCCCCAACGACCCCTGGGTTGTCCTCTCAGAG

ATGGCTATGGATACTACAAGGTGTGGAGCCCAGTTGTTGACTCTGGTCGAGCAGATCCTGGCAGAGTTCCAGCTG
Met Ala Met Asp Thr Thr Arg Cys Gly Ala Gln Leu Leu Thr *Leu* Val Glu Gln Ile Leu Ala Glu Phe Gln Leu 25

CAGGAGGAAGACCTGAAGAAGGTGATGAGCCGGATGCAGAAGGAGATGGACCGTGGCCTGAGGCTGGAGACCCAC
Gln Glu Glu Asp Leu Lys Lys Val Met Ser Arg Met Gln Lys Glu Met Asp Arg Gly Leu Arg Leu Glu Thr His 50

GAGGAGGCCAGTGTAAAGATGTTACCCACCTACGTGCGTTCCACCCCAGAAGGCTCAGAAGTCGGAGACTTTCTC
Glu Glu Ala Ser Val Lys Met Leu Pro Thr Tyr Val Arg Ser Thr Pro Glu Gly Ser *Glu* Val Gly Asp Phe Leu 75

TCCTTAGACCTGGGAGGAACCAACTTCAGAGTGATGCTGGTCAAAGTGGGAGAGGGGGAGGCAGGGCAGTGGAGC
Ser Leu <u>Asp</u> <u>Leu</u> <u>Gly</u> <u>Gly</u> <u>Thr</u> Asn Phe Arg Val Met Leu Val Lys Val Gly Glu Gly Glu Ala Gly **Gln** Trp Ser 100

GTGAAGACAAAACACCAGATGTACTCCATCCCCGAGGACGCCATGACGGGCACTGCCGAGATGCTCTTTGACTAC
Val Lys Thr Lys His Gln Met Tyr Ser Ile Pro Glu Asp Ala Met Thr Gly Thr Ala Glu *Met* Leu Phe Asp Tyr 125

ATCTCTGAATGCATCTCTGACTTCCTTGACAAGCATCAGATGAAGCACAAGAAACTGCCCCTGGGCTTCACCTTC
Ile Ser Glu Cys Ile Ser Asp Phe Leu Asp Lys His Gln Met Lys His Lys Lys Leu Pro Leu Gly Phe Thr Phe 150

TCCTTCCCTGTGAGGCACGAAGACCTAGACAAGGGCATCCTCCTCAATTGGACCAAGGGCTTCAAGGCCTCTGGA
SER Phe Pro Val Arg His Glu Asp Leu Asp *Lys* Gly Ile Leu Leu Asn Trp Thr Lys Gly Phe Lys Ala Ser **Gly** 175

GCAGAAGGGAACAACATCGTAGGACTTCTCCGAGATGCTATCAAGAGGAGAGGGGACTTTGAGATGGATGTGGTG
Ala Glu Gly Asn Asn Ile **Val** Gly Leu Leu **Arg** Asp Ala Ile Lys Arg Arg *Gly* Asp Phe Glu Met Asp Val Val 200

GCAATGGTGAACGACACAGTGGCCACAATGATCTCCTGCTACTATGAAGACCGCCAATGTGAGGTCGGCATGATT
Ala Met **Val** ASN <u>ASP</u> Thr Val Ala Thr Met Ile Ser Cys Tyr Tyr Glu Asp Arg Gln Cys Glu Val Gly Met Ile 225

GTGGGCACTGGCTGCAATGCCTGCTACATGGAGGAAATGCAGAATGTGGAGCTGGTGGAAGGGGATGAGGGACGC
Val *Gly* <u>**Thr**</u> Gly Cys Asn Ala Cys Tyr Met Glu Glu Met Gln Asn Val Glu Leu Val Glu Gly Asp Glu Gly Arg 250

ATGTGCGTCAACACGGAGTGGGGCGCCTTCGGGGACTCGGGCGAGCTGGATGAGTTCCTACTGGAGTATGACCGG
Met Cys Val Asn Thr **GLU** Trp Gly Ala Phe **Gly** Asp Ser Gly Glu Leu Asp Glu Phe Leu Leu Glu Tyr Asp Arg 275

ATGGTGGATGAAAGCTCAGCGAACCCCGGTCAGCAGCTGTACGAGAAGATCATCGGTGGGAAGTATATGGGCGAG
Met Val Asp **Glu** Ser Ser Ala Asn Pro Gly Gln Gln *Leu* Tyr GLU Lys Ile Ile Gly Gly Lys Tyr Met **Gly** **Glu** 300

CTGGTACGACTTGTGCTGCTTAAGCTGGTGGACGAGAACCTTCTGTTCCACGGAGAGGCCTCGGAGCAGCTGCGC
Leu Val Arg Leu Val Leu Leu Lys **Leu** Val Asp Glu Asn Leu Leu Phe His Gly Glu Ala Ser Glu Gln Leu Arg 325

[illegible]TCCTTTTCAC[illegible]GTGTCACAAGTGGAGAGCGACTCCGGGGACCGAAAGCAGATCCACAAC
Thr Arg Gly Ala Phe Glu Thr Arg Phe Val Ser [illegible] Gln Ile His Asn 350

ATCCTAAGCACTCTGGGGCTTCGACCCTCTGTCACCGACTGCGACATTGTGCGCCGTGCCTGTGAAAGCGTGTCC
Ile Leu Ser Thr Leu Gly Leu Arg Pro Ser Val Thr Asp Cys Asp Ile Val Arg Arg Ala Cys Glu Ser Val Ser 375

ACTCGCGCCGCCCATATGTGCTCCGCAGGACTAGCTGGGGTCATAAATCGCATGCGCGAAAGCCGCAGTGAGGAC
Thr Arg Ala Ala His Met Cys Ser Ala Gly Leu Ala Gly Val Ile Asn Arg Met Arg Glu Ser Arg Ser Glu Asp 400

GTGATGCGCATCACTGTGGGCGTGGATGGCTCCGTGTACAAGCTGCACCCGAGCTTCAAGGAGCGGTTTCACGCC
Val Met Arg Ile Thr Val Gly Val <u>Asp</u> <u>Gly</u> <u>Ser</u> <u>Val</u> <u>Tyr</u> Lys Leu His Pro *Ser* Phe Lys Glu Arg Phe His Ala 425

AGTGTGCGCAGGCTGACACCCAACTGCGAAATCACCTTCATCGAATCAGAGGAGGGCAGCGGCAGGGGAGCCGCA
Ser Val Arg Arg Leu Thr Pro Asn Cys Glu Ile Thr Phe Ile Glu Ser Glu Glu Gly Ser Gly Arg Gly Ala Ala 450

CTGGTCTCTGCGGTGGCCTGCAAGAAGGCTTGCATGCTGGCCCAGTGAAATCCAGGTCATATGGACCGGGACCTG
Leu Val Ser Ala Val Ala Cys Lys Lys Ala Cys Met Leu Ala Gln 465

GGTTCCACGGGGACTCCACACACCACAAATGCTCCCAGCCCACCGGGGCAGGAGACCTATTCTGCTGCTACCCCT

GGAAAATGGGGAGAGGCCCCTGCAAGCCGAGTCGGCCAGTGGGACAGCCCTAGGGCTCTCAGCCTGGGGCAGGGG

GCTGGGAGGAAGAAGAGGATCAGAGGCGCCAAGGCCTTTCTTGCTAGAATCAACTACAGAAAATGGCGGAAAATA

CTCAGGACTTGCACTTTCACGATTCTTGCTTCCCAAGCGTGGGTCTGGCCTCCCAAGGGAATGCTTCCTGGACCT

TGCAATGGCCTGGCTTCCCTGGGGGGGACACACCTTCATGGGGAGGTAACTTCAGCAGTTCGGCCAGACCAGACC

CCAGGAGAGTAAGGGCTGCTGGTCACCCAGACCTGGCTGTTTTCTTGTCTGTGGCTGAAGAGGCCGGGGAGCCAT

GAGAGACTGACTATCCGGCTACATGGAGAGGACTTTCCAGGCATGAACATGCCAGAGACTGTTGCCTTCATATAC

CTCCACCCGAGTGGCTTACAGTTCTGGGATGAACCCTCCCAGGAGATGCCAGAGGTTAGAGCCCCAGAGTCCTTG

CTCTAAGGGGACCAGAAAGGGGAGGCCTCACTCTGCACTATTCAAGCAGGAATCATCTCCAACACTCAGGTCCCT

GACCCAGGAGGAAGAAGCCACCCTCAGTGTCCCTCCAAGAGACCACCCAGGTCCTTCTCTCCCTCGTTCCCAAAT

GCCAGCCTCTCTACCTGGGACTGTGGGGGAGTTTTTAATTAAATATTTAAAACTT (A_n)

Fig. 5.2. DNA and amino acid sequences of hexokinase D from rat liver. The last residue in each exon is italicized; residues linked to MODY (maturity-onset diabetes of the young) are shown in ***bold****; residues involved in glucose binding are shown in CAPITALS; and residues involved in ATP binding are* <u>*underlined*</u>.

was different because of different promoter regions in the two tissues. References 2 and 56 are recommended for general information about the material in this section.

ALTERNATIVE PROMOTERS FOR THE HEXOKINASE D GENE

After the cloning of cDNA for hepatic hexokinase D, Iynedjian and colleagues obtained an important piece of information. They showed not only that hexokinase D mRNA was present only in liver and islets of Langerhans, as expected, but also that the islet cDNA was bigger than that of liver: 2.8 kb instead of 2.4 kb.[19] It was thus 400 nucleotides longer, as was the one from tumor islet cells. This result was surprising in view of the demonstration of only one gene for the enzyme. The islet mRNA could hybridize with several DNA probes covering various regions of the liver mRNA; confirming that both proteins were coded by the same gene, and indicating that the regions of sequence identity were distributed over most of the length of the mRNA. There was, however, a region that failed to hybridize, that of the 5' untranslated leader of the liver mRNA: the complementary probe failed to hybridize with RIN cell mRNA, pointing to a sequence difference in the 5' area of the two mRNAs. This difference was localized in a region upstream of the codon for amino acid 19, and was attributed by the authors to a tissue-specific differential splicing, or the use of alternative promoters, or both. Multiple promoters had been described for a number of eukaryotic genes,[57,58] so this appeared quite plausible. In some of these cases, the choice among alternative promoters is specific to particular cell types or developmental stages and is accompanied by alternative splicing, so that transcripts initiated at the upstream promoter in one tissue subsequently undergo processing with removal of the leader sequence found in the mRNA transcribed from a downstream promoter in another tissue or at another stage of development.[59] Iynedjian speculated that a similar mechanism might apply to the hexokinase D gene.[19]

Iynedjian's initial suggestion of the existence of alternative promoters and differential splicing of the hexokinase D gene proves to be correct. The structure of the hexokinase D gene, which in rat is 15.5 kb long and consists of 10 exons separated by 9 introns, appears in Figure 5.3, which is based on information in Figure 1 of his review.[2] It shows alternative promoters responsible for the initiation of transcription at different sites on the DNA in hepatic and endocrine cells. The first clue to the existence of promoters specific to cell types came from the sequences of two quasi-full-length cDNAs isolated from ratliver and insulinoma libraries.[47,50] The sequences were essentially identical for more than 2000 nucleotides starting from the 3' ends of the cDNAs, but segments of approximately 100 nucleotides at the 5' ends were found to be entirely different. The 5' specific sequences of the cDNAs

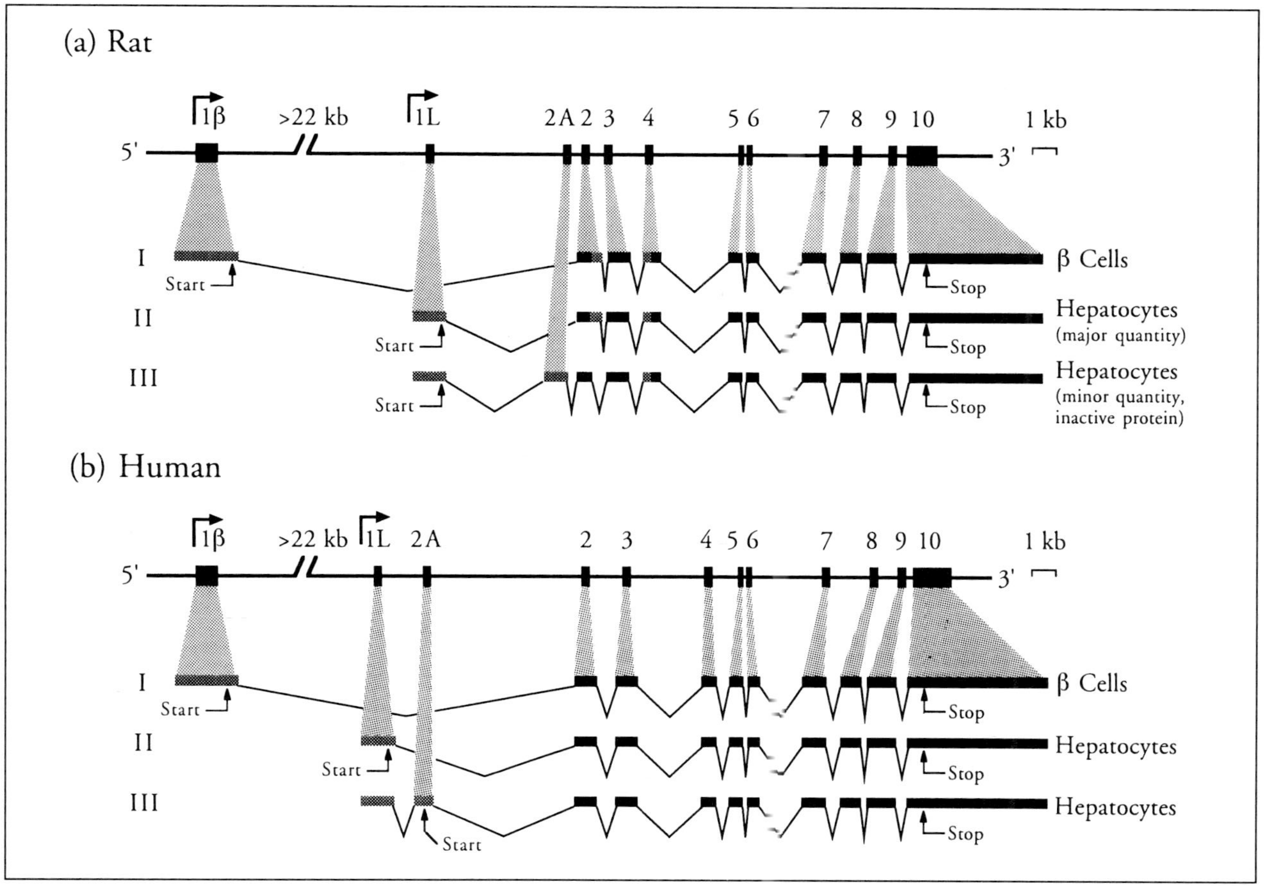

Fig. 5.3. Structure of the hexokinase D gene in (a) rat and (b) human. The top lines show the exon-intron organization, with exons represented by the black boxes, and transcription initiation sites by arrows (β, β-cell; L, hepatocyte). Below are shown in expanded views the variant mRNAs produced in β-cells (I) and hepatocytes (II and III) by tissue-specific initiation of transcription and alternative splicing; in these lines alternatively spliced exons or parts of exons are shaded, and the initiation and termination codons are marked with arrows. In the rat the mRNA of line III is quantitatively minor and codes an inactive protein, but it is not known whether this is also true for the human. Based on Figure 1 of Biochem J 1993; 293:1-13.

were mapped to widely separated sites in genomic DNA by Magnuson and coworkers.[50,54] They further identified nine exons, numbered 2-10 in the transcription unit, whose assembly gives rise to the common sequence found in the liver and insulinoma-derived cDNAs. The leader exon encoding the 5' end of the hepatic mRNA is termed exon 1L. It was contained in a phage λ clone which also carried the common exons 2-4. The leader exon for the 5' end of the insulinoma mRNA, termed exon 1β in reference to β-cells of the islets of Langerhans, was localized in a different phage clone with non-overlapping genomic DNA. It was therefore concluded that the liver-specific exon 1L was contiguous to the body of the structural gene, whereas the islet-type exon 1β was located at an unspecified distance further upstream. More than 22 kb of DNA separate the two leader exons in the rat gene.

The fact that the upstream exon 1β is used both in pancreatic β cells and in pituitary cells and the downstream exon 1L exclusively in liver was established by primer extension and nuclease protection experiments and further confirmed by reverse transcription and polymerase chain reaction.[50] It should be noted that the two tissue-specific exons 1 of the hexokinase D gene specify not only the 5' untranslated regions of the islet and liver mRNAs, but also their initial 45 nucle-

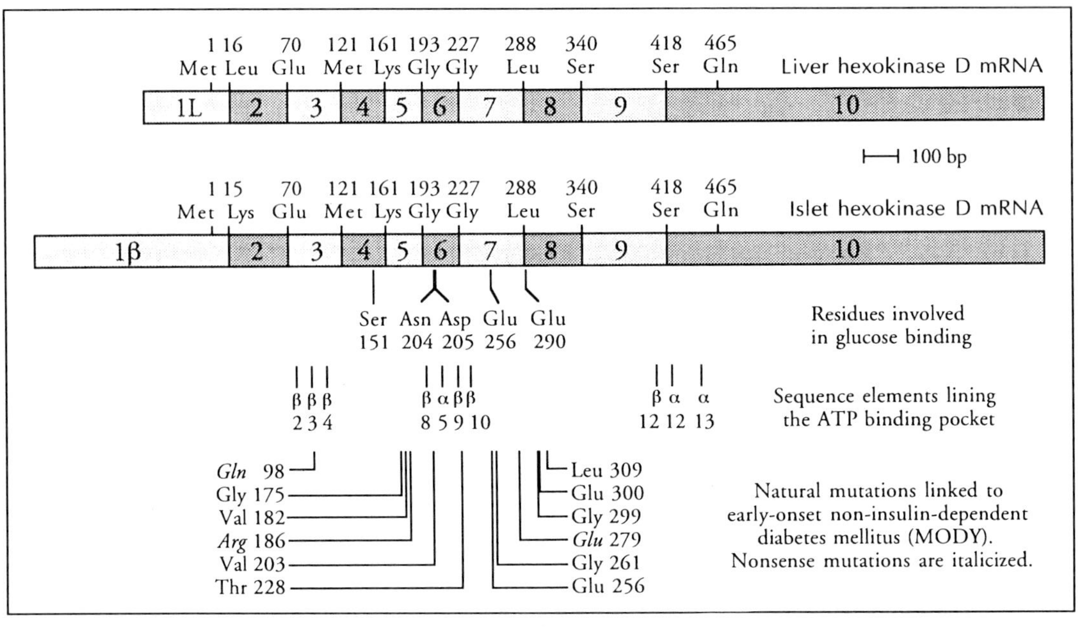

Fig. 5.4. Functional assignments of amino acid residues in hexokinase D. The top lines show the exon structures of the mRNAs of hepatocyte and β-cell hexokinase D. In each case the first and last amino acid residues of the proteins are marked, as well as the last residue coded by each exon. The lower part of the diagram shows the residues known to be involved in glucose binding, the sections of α-helix (α) or β-sheet (β) lining the ATP binding pocket, and residues that have been found to have undergone mutations in maturity-onset diabetes of the young (MODY). Adapted from Figure 3 of Biochem J 1993; 293:1-13.

otides of protein coding sequence. It follows that distinct hexokinase D isoforms are produced in the liver and pancreatic islets, which differ in primary structure by 15 amino acids (including the initiator Met) at the *N*-terminal ends of the molecules, out of a total sequence of 465 amino acid residues.

As indicated in Figure 5.4, which is based on Figure 3 of Iynedjian,[2] comparison of the hexokinase D sequence with that of yeast hexokinase has allowed residues involved with substrate binding to be identified, as well as ones linked with maturity-onset diabetes of the young (MODY), which will be discussed shortly.[60]

ALTERNATIVE MRNA SPLICING WITHIN EACH TISSUE: HEXOKINASE D ISOFORMS

In addition to the differential splicing of leader exons associated with the cell-specific control of transcription initiation, other modes of alternative splicing are known to affect hexokinase D transcripts. Thus there is alternative splicing of hexokinase D mRNA within the liver itself.[61] An optional cassette exon has been identified in the rat gene between the originally described exons 1L and 2, 370 bases upstream from exon 2 (Fig. 5.3). This cassette exon of 151 base pairs, termed exon 2A, is retained in a minor fraction (about 5%) of hexokinase D mRNA in rat liver.[61] Alternative donor sites in exon 2 are used with low frequency and lead to deletions of either 52 or 25 nucleotides from mature mRNA. Finally an acceptor site in exon 4 located 51 nucleotides downstream of the major splice site has been described.[49,50,54]

This alternative splicing removes 51 bp of coding sequence, causing a 17 amino acid in-frame deletion from a portion of the enzyme between the putative glucose and ATP binding domains.[50,54] Magnuson has designated the major liver isoform of hexokinase D as L1 and the major islet isoform as B1.[62] The variant islet isoform resulting from the alternative splice acceptor site in the fourth exon of the gene has been termed B2 and the corresponding liver isoform L2. These minor isoforms B2 and L2 are immunoreactive but they lack catalytic activity.[49] It appears therefore that deletion of 17 amino acids between the putative glucose and ATP binding domains is severely detrimental to the expression of hexokinase D activity.

Alternative RNA splicing appears to be a means of generating isoform diversity and of regulating gene expression.[63] However, as both B2 and L2 isoforms are functionally inactive (at least as hexokinase D) and do not appear to accumulate when transfected into NIH-3T3 cells it seems improbable that either has a direct functional role. However, the question is whether splice site selection is a regulated or a constitutive process in the liver and the islet. As mentioned in Chapter 4, hexokinase D mRNA is also expressed in the pituitary gland, using the upstream promoter, but as a result of two alternative splices there is no activity there, in agreement with the idea that the different modes of splicing are responsible for the formation of minor mRNA species that encode enzymically inactive polypeptides.

THE HUMAN HEXOKINASE D GENE

Reports on the cloning of the human hexokinase D gene demonstrate that the major structural features of the rat gene have been conserved in the human species.[53,55] It has the same exon/intron organization apart from a third alternative exon downstream from exon 1β. Most important, the dual promoter arrangement with upstream islet-specific exon 1β and downstream liver-specific exon 1L is very similar in the human and rat transcription units (Fig 5.3b). As in the rat, the mature hexokinase D mRNAs of human islets and liver are assembled from cell-type specific exons 1 and common exons 2-10.[53,64] The cell-specific exons harbor the initiator ATG triplets, which are located 45 and 48 nucleotides upstream from the donor splice sites in exons 1β and 1L respectively. Reminiscent of the situation in the rat, an alternative transcript in human liver is produced by splicing of an optional cassette exon 2a between exons 1L and 2. In this case, the open reading frame for the protein begins at an ATG codon situated 42 nucleotides upstream from the donor splice site of the cassette exon, predicting a variant *N*-terminal domain of 14 amino acid residues in the corresponding protein.[64] The gene has been mapped to human chromosome 7p.[51,65]

RELATIONSHIP BETWEEN THE HEXOKINASE D GENE AND NIDDM

Fasting hyperglycemia, the hallmark of diabetes mellitus, is largely the result of glucose overproduction by the liver. Hepatic glucose production is the balance between the fluxes through hexokinase D and glucose-6-phosphatase, as discussed in Chapter 4. As we saw in Chapter 1, hexokinase D activity is markedly decreased in insulin-deficient diabetic rats, whereas glucose-6-phosphatase activity is markedly increased. Changes in the circulating insulin concentrations consequently affect hepatic glucose flux.

Not all forms of diabetes are dependent on insulin, however. There is not only diabetes type 1, or insulin-dependent diabetes mellitus (IDDM), whose name reflects the need for insulin replacement therapy to prevent the pathology derived from insulin deficiency due to a dysfunction of the β cells; there is also diabetes type 2 or non-insulin-dependent diabetes mellitus (NIDDM), in which the hyperglycemia is not the consequence of β-cell destruction, but a failure of seemingly normal β-cells to meet an increased demand for insulin. A common feature of the these two hyperglycemic syndromes is the absence of glucose-stimulated insulin secretion, but whereas the onset of IDDM occurs in childhood and young adulthood, that of NIDDM occurs normally in middle or old age. Recent DNA polymorphism studies have established a linkage between the hexokinase D locus and a particular form of diabetes termed maturity-onset diabetes of the young (MODY) a subtype of non-insulin-dependent diabetes mellitus, also often known as early-onset NIDDM, as it presents from the second decade. (The terminology is confusing; an age that is regarded as "mature" for diabetes mellitus type 1 is considered "early" for NIDDM). Additional information about the topics in this section may be found in References 3 and 5, 66, 67, and 68.

NIDDM AND MODY

NIDDM is a common metabolic disorder characterized by impaired insulin secretion in response to glucose and resistance of the target tissues to the action of insulin, including exaggerated hepatic glucose output in the postabsorptive state. Both defects contribute to the development of fasting hyperglycemia at full-blown stages of the disease. The etiology of NIDDM is not understood in detail, but it is known to involve both genetic and environmental factors. As a general rule, the disease is classified among the multifactorial polygenic syndromes. It is a major public health problem that affects 5-7% of the world population. Although most forms of NIDDM do not exhibit simple Mendelian inheritance, the contribution of heredity to its development has been recognized for many years.

The mild form of NIDDM called MODY is characterized by an early age of onset and Mendelian autosomal dominant transmission with high penetrance. The fact that MODY appears to have a simple autosomal dominant mode of inheritance made it attractive in the search for disease genes linked to NIDDM. A problem, however, is the diagnosis; the age of onset is really a poorly defined and qualitative criterion, as the early stages of the disease are often asymptomatic and only disclosed when younger relations of NIDDM patients are tested. Furthermore, some cases of early-onset NIDDM may be phenocopies of MODY, with two diabetic parents contributing multiple diabetes susceptibility genes to their offspring.[66] According to the World Health Organization (WHO) the criterion for clinical diabetes is a plasma glucose concentration higher than 7.8 mM (140 mg/cl) 2h after an oral glucose load (glucose intolerance according to WHO) or a fasting plasma glucose higher than 6.1 mM (110 mg/cl) in two separate measurements.

A good definition of MODY and delineation of its relationship to NIDDM comes from studies of more than 200 French families,[67] which were examined in a campaign to identify multigenerational pedigrees with NIDDM. Of these, 32 satisfied the criteria for MODY, defined in this work as the same as the WHO criterion for clinical diabetes, with the additional requirement that at least two individuals were diagnosed before 25 years of age. With these moderately stringent criteria for diagnosis, MODY may account for as much as 10-15% of familial NIDDM.

LINKAGE BETWEEN THE HEXOKINASE D GENE AND MODY

Although clinical and metabolic profiles of families with maturity-onset diabetes of the young are diverse, most MODY patients exhibit a decreased insulin response to glucose and hyperglycemia, suggesting a primary pancreatic β-cell defect.[68] Several genes whose products seem to be involved in insulin secretion, such as liver or pancreatic β-cell glucose transporter (GLUT2) and hexokinase D, are candidate genes for MODY. The hexokinase D regulatory protein (Chapter 7) is an obvious additional candidate gene.

Two separate studies have reported linkage of the MODY phenotype to the hexokinase D locus in a number of multigenerational pedigrees with the disease.[67,69] These strongly suggest that mutations in the hexokinase D locus on chromosome 7p might predispose individuals in these families to MODY.

The transmission of hexokinase D alleles was followed using two polymorphic microsatellite DNA markers in the immediate vicinity of the gene.[65,67] Specific mutations have been identified in MODY patients in several of the above families.[55,70,71] The mutations were not found in unaffected relatives or in control subjects in the general population, suggesting their critical role in the disease. The reported mutations include nonsense, missense and splicing-site mutations, and the positions of the amino acid residues affected by nonsense or missense mutations are shown in Figure 5.4. The report of a nonsense mutation in exon 7 of the hexokinase D gene and its linkage with MODY was the first evidence implicating a mutation in a gene involved in glucose metabolism in the pathogenesis of NIDDM;[70] shortly afterwards a missense mutation was detected.[71] To assess the effect of the missense mutations on the enzymic activity of hexokinase D, Gidh-Jain and coworkers reproduced these mutations by in vitro mutagenesis of a human islet cDNA and analyzed the kinetic properties of the resulting proteins after bacterial expression.[60] The amino acid substitutions were shown to have more or less pronounced deleterious effects on the V_{max} or the glucose affinity of the enzyme or both. The same approach has been used to analyze other mutants,[6] and Chapter 3 includes further discussion of these experiments.

As MODY has a dominant transmission with high penetrance it seems odd at first sight that it could be due to a fault in an enzyme such as hexokinase D; one would expect this to lead to a recessive transmission, as metabolic control analysis predicts that virtually all such genetic lesions will be recessive.[72] However, when it is remembered that MODY is symptomless, i.e. its definition does not imply a qualitative characteristic but a quantitative one, glycemia higher than a specified value, the relationship becomes more comprehensible. When the first mutation in the hexokinase D gene and the linkage with MODY were reported it was suggested that mutations in the hexokinase D gene lead to the development of glucose intolerance by a gene-dose mechanism.[71] The demonstration of decreased enzyme activity of missense mutations in hexokinase D associated with MODY[60] is consistent with this hypothesis. Decreased cellular levels of hexokinase D activity may increase the glucose threshold for insulin secretion in the β cell as well as impair hepatic uptake and metabolism of glucose. As diabetic patients have a normal as well as a mutant hexokinase D gene[70,71] they are not totally deficient, but rather are predicted to have hexokinase D levels about 50% of normal. However, this level of activity is apparently insufficient to ensure normal β-cell and/or hepatocyte function, as these individuals have readily detectable defects in glucose metabolism that include mild fasting hyperglycemia and impaired glucose tolerance.

Mutations in the hexokinase D gene have been found not only in French families with MODY but also in British, Swedish and African-American patients, and thus appear not to be restricted to a particular ethnic group. On the other hand the hexokinase D gene appears not to be implicated in the development of late-onset NIDDM, as linkage between hexokinase D and NIDDM has been definitely excluded in 79 multigenerational French families.[73] Similar results have been obtained with Welsh and English families,[74,75] and in 18 pedigrees from Utah with multiple type 2 diabetic individuals.[76]

What are the clinical consequences of mutations in the hexokinase D gene? Velho and coworkers evaluated patients from four MODY kindreds with four distinct hexokinase D mutations for islet β-cell function.[77] The patients showed a normal first-phase elevation of plasma insulin in response to injection of glucose, but failed to sustain adequate insulin levels during a continuous glucose infusion set to maintain the plasma glucose concentration at 10 mM. In the basal state these subjects also displayed inappropriately low plasma levels of insulin in relation to the slightly elevated blood glucose levels. Together, data from these patients are compatible with an increase in the glucose threshold for insulin secretion. Such a defect is a predicted consequence of a reduction in β-cell hexokinase D activity, as shown in a theoretical model of glucose-stimulated insulin secretion proposed several years ago by Meglasson and Matschinsky.[1]

Investigations of hepatic glucose metabolism in subjects with MODY have not yet been reported. At our current state of knowledge the MODY syndrome associated with mutations of hexokinase D emerges as a disorder characterized by mild hyperglycemia appearing during childhood, resulting primarily from a discrete defect in glucose-induced insulin secretion.[71,77] In that respect it differs from more prevalent forms of NIDDM in which the primary abnormality appears to be the resistance of the liver and peripheral tissues to the action of insulin.[78]

PERSPECTIVES

Without doubt, transgenic mice offer great opportunities for solving many problems of metabolism. That is also the case with ribozymes, which can be expressed in transgenic mice; these are RNA molecules that possess catalytic RNA-cleavage activity,[79,80] an activity that can be made specific by flanking the ribozyme catalytic element with two gene-specific fragments in antisense orientation. In this way the ribozyme activity can be targeted against a unique RNA sequence, thus reducing target mRNA level and activity.

A recent and very interesting article describes the use of this strategy to generate a mouse model for MODY in which the hexokinase D activity was specifically reduced in β-cells, but not in hepatocytes.[42] The specific reduction was achieved by expressing a hexokinase D ribozyme in β-cells in transgenic mice, which manifest only a third of the normal islet hexokinase D activity. As expected from the proposed model for insulin secretion,[38] insulin release from pancreas perfused in situ was impaired in response to glucose, in a manner similar to that observed in MODY patients. This finding that partial attenuation of expression of the normal hexokinase D protein is sufficient to impair the sensitivity of β-cells to glucose supports the interpretation of the dominance of the hexokinase D mutations in MODY as a gene-dosage effect. However, in spite of the considerable reduction of hexokinase D activity in β-cells, to below the level that gives rise to diabetes in MODY patients, the transgenic mice remain euglycemic, i.e. their levels of glucose and insulin in the plasma remained normal.

The authors give two alternative interpretations to explain their results.[42] It is possible that in the mouse other insulin secretagogues can compensate for the reduction in glucose-induced secretion. Alternatively, an impaired liver function, in addition to that of β-cells, is required for the induction of diabetes by hexokinase D deficiency. As a matter of fact, such liver impairments have been documented in MODY patients.[81]

In view of the great complexity of the mechanisms involved in insulin secretion as outlined in Figure 5.1, and the participation of neurotransmitters and hormones in its modulation, it is not surprising that the response of the isolated islet should differ from that in the whole animal. The idea that impairment in liver hexokinase D is also required is quite possible, because the fact that the mutations found in MODY patients are in common exons means that they must affect both pancreatic and liver forms of the enzyme, and consequently both forms should be impaired in these patients. More experiments will be needed to illuminate this emerging field, and the availability of transgenic mice and ribozymes will provide a valuable tool.

REFERENCES

1. Meglasson MD, Matschinsky FM. New perspectives on pancreatic islet glucokinase. Am J Physiol 1984; 246; E1-E13.
2. Iynedjian PB. Mammalian glucokinase and its gene. Biochem J 1993; 293:1-13.
3. Permutt MA, Chiu KC, Tanizawa Y. Glucokinase and NIDDM. A candidate gene that paid off. Diabetes 1992; 41:1367-1372.
4. Matschinsky FM. Glucokinase as glucose sensor and metabolic signal generator in pancreatic β cells and hepatocytes. Diabetes 1990; 39:647-652.
5. Randle PJ. Glucokinase and candidate genes for type 2 (non-insulin-dependent) diabetes mellitus. Diabetologia 1993; 36:269-275.
6. Pilkis SJ, Weber IT, Harrison RW et al. Glucokinase: structural analysis of a protein involved in susceptibility to diabetes. J Biol Chem 1994; 269:21925-21928.
7. Coore HG, Randle PJ. Regulation of insulin secretion studied with pieces of rabbit pancreas incubated in vitro. Biochem J 1964; 93:66-78.
8. Matschinsky FM, Ellerman JE. Metabolism of glucose in the islets of Langerhans. J Biol Chem 1968; 243:2730-2736.
9. Ashcroft SJ, Randle PJ. Enzymes of glucose metabolism in normal mouse pancreatic islets. Biochem J 1970; 119:5-12.
10. Hedeskov CJ, Capito K. The effect of starvation on insulin secretion and glucose metabolism in mouse pancreatic islets. Biochem J 1974; 140:423-433.
11. Malaisse WJ, Sener A, Levy J. The stimulus-secretion coupling of glucose-induced insulin release: fasting-induced adaptation of key glycolytic enzymes in pancreatic islets. J Biol Chem 1976; 251:1731-1737.
12. Täljedal IB. On insulin secretion. Diabetologia 1981; 21:1-17.
13. Ashcroft SJH. Glucoreceptor mechanisms and control of insulin release biosynthesis. Diabetologia 1980; 18:5-15.
14. Trus MD, Zawalich KC, Burch PT et al. Regulation of glucose metabolism in pancreatic islets. Diabetes 1981; 30:911-922.
15. Burch PT, Trus M, Berner DK et al. Adaptation of glycolytic enzymes. Glucose use and insulin release in rat pancreatic islets during fasting and refeeding. Diabetes 1981; 30:923-928.
16. Meglasson MD, Burch PT, Berner DK et al. Chromatographic resolution and kinetic characterization of glucokinase from islets of Langerhans. Proc Natl Acad Sci USA 1983; 80:85-89.
17. Meglasson MD, Burch PT, Hoenig M et al. Identification and significance of glucokinase in transplantable insulinomas. J Biol Chem 1983; 258:2094-2097.
18. Iynedjian PB, Möbius G, Seitz HJ et al. Tissue-specific expression of glucokinase: identification of the gene product in liver and pancreatic islets. Proc Natl Acad Sci USA 1986; 83:1998-2001.
19. Iynedjian PB, Pilot P-R, Nouspikel T et al. Differential expression and regulation of the glucokinase gene in liver and islets of Langerhans. Proc Natl Acad Sci USA 1989; 86:7838-7842.

20. Jetton TL, Magnuson MA. Heterogenous expression of glucokinase among pancreatic β-cells. Proc Natl Acad Sci USA 1992; 89:2619-2623.
21. Meglasson MD, Schinco M, Matschinsky FM. Mannose phosphorylation by glucokinase from liver and transplantable insulinoma: cooperativity and discrimination of anomers. Diabetes 1983; 32:1146-1151.
22. Meglasson MD, Matschinsky FM. Discrimination of glucose anomers by glucokinase from liver and transplantable insulinoma. J Biol Chem 1983; 258:6705-6708.
23. Vischer U, Blondel B, Wollheim CB et al. Hexokinase isoenzymes of RIN-m5F insulinoma cells: expression of glucokinase gene in insulin-producing cells. Biochem J 1987; 241:249-255.
24. Ashcroft SJH, Hedeskov CJ, Randle PJ. Glucose metabolism in mouse pancreatic islets. Biochem J 1970; 118:143-154.
25. Sener A, Levy J, Malaisse WJ. The stimulus-secretion coupling of glucose-induced insulin release. Does glycolysis control calcium transport in the B-cell? Biochem J 1976; 156:521-525.
26. Zawalich WS, Matschinsky FM. sequential analysis of the releasing and fuel functions of glucose in isolated perifused pancreatic islets. Endocrinology 1977; 100:1-8.
27. Ashcroft SJH, Weerasinghe LCC, Basset JM et al. The pentose cycle and insulin release in mouse pancreatic islets Biochem J 1972; 126:525-532.
28. Grodsky GM. The kinetics of insulin release. In Insulin II. (Hasselblatt A, Bruchhausen FV, eds.), New York: Springer Verlag, 1975:1-16.
29. Schuit RC, In'tVeld PA, Pipeleers DG. Glucose stimulates proinsulin biosynthesis by a dose-dependent recruitment of pancreatic beta cells. Proc Natl Acad Sci USA 1988; 85:3865-3869.
30. Docherty K, Clark AR. Nutrient regulation of insulin gene expression. FASEB J. 1994; 8:20-27.
31. Hellman B, Sehlin J, Täljedal IB. Evidence for mediated transport of glucose in mammalian pancreatic β-cells. Biochim Biophys Acta 1971; 241:147-154.
32. Thorens B, Sarkar HK, Kaback HR et al. Cloning and functional expression in bacteria of a novel glucose transporter present in liver, intestine, kidney and β-pancreatic islet cells. Cell 1988; 55:281-290.
33. Johnson JH, Newgard CB, Milburn JL et al. The high K_m glucose transporter of islets of Langerhans is functionally similar to the low affinity transporter of liver and has an identical primary sequence. J Biol Chem 1990; 265:6548-6551.
34. Pessin JE, Bell GI. Mammalian facilitative glucose transporter family: structure and molecular regulation. Annu Rev Physiol 1992; 54:911-930.
35. Tal M, Liang Y, Najali H, Lodish HF et al. Expression and function of GLUT-1 and GLUT-2 glucose transporter isoforms in cells of cultured rat pancreatic islets. J Biol Chem 1992; 267:17241-17247.
36. Efrat S, Tal M, Lodish HF. The pancreatic β-cell glucose sensor. Trends Biochem Sci 1994; 19:535-538.
37. Giroix M-H, Sener A, Pipeleers DG et al. Hexose metabolism in pancre-

atic islets. Inhibition of hexokinase. Biochem J 1984; 223:447-453.
38. Meglasson MD, Matschinsky FM. Pancreatic islet glucose metabolism and regulation of insulin secretion. Diabetes/Metabolism Revs. 1986; 2:163-214.
39. Ashcroft FM, Harrison DE, Ashcroft SJH. Glucose induces closure of single potassium channels in isolated pancreatic β-cells. Nature, 1984; 312:446-448.
40. Cook DL, Hales CN. Intracellular ATP directly blocks K^+ channels in pancreatic B-cells. Nature 1984; 311:271-273.
41. Brown GC. Control of respiration and ATP synthesis in mammalian mitochondria and cells. Biochem J 1992; 284:1-13.
42. Efrat S, Leiser M, Wu Y-J et al. Ribozyme-mediated attenuation of pancreatic β-cell glucokinase expression in transgenic mice results in impaired glucose-induced insulin secretion. Proc Natl Acad Sci USA 1994; 91:2051-2055.
43. Waddell ID, Burchell A. The microsomal glucose 6-phosphatase enzyme of pancreatic islets. Biochem J 1988; 255:471-476.
44. Khan A, Chandramouli V, Östenson C-G et al. Evidence for the presence of glucose cycling in pancreatic islets of the ob/ob mouse. J Biol Chem 1989; 264:9732-9733.
45. Khan A, Chandramouli V, Ostensson C-G et al. Glucose cycling in islets from healthy and diabetic rats. Diabetes 1990; 39:456-459.
46. Täljedal IB. Presence, induction and possible role of glucose 6-phosphatase in mammalian pancreatic islets. Biochem J 1969; 114:387-394.
47. Andreone TL, Printz R.L, Pilkis SJ et al. The amino acid sequence of rat liver glucokinase deduced from cloned cDNA. J Biol Chem 1989; 264:363-369.
48. Hughes SD, Quaade C, Milburn JL et al. Expression of normal and novel glucokinase mRNAs in anterior pituitary and islet cells. J Biol Chem 1991; 266:4521-4530.
49. Liang Y, Jetton TL, Zimmerman EC et al. Effects of alternate RNA splicing on glucokinase isoform activities in the pancreatic islet, liver and pituitary. J Biol Chem 1991; 266:6999-7007.
50. Magnuson MA, Shelton KD. An alternate promoter in the glucokinase gene is active in the pancreatic β-cell. J Biol Chem 1989; 264:15936-15942.
51. Nishi S, Stoffel M, Xiang K et al. Human pancreatic beta-cell glucokinase: cDNA sequence and localization of the polymorphic gene to chromosome 7, band p 13. Diabetologia, 1992; 35:743-747.
52. Koranyi L, Tanizawa Y, Welling CM et al. Human islet glucokinase gene. Isolation and sequence analysis of full-length cDNA. Diabetes 1992; 41:807-811.
53. Tanizawa Y, Koranyi LI, Welling CM et al. Human liver glucokinase gene: cloning and sequence determination of two alternatively spliced cDNAs. Proc Natl Acad Sci USA 1991; 88:7294-7297.
54. Magnuson MA, Andreone TL, Printz RL et al. Rat glucokinase gene: structure and regulation by insulin. Proc Natl Acad Sci USA. 1989; 86:4838-4842.
55. Stoffel M, Froguel P, Takeda J et al. Human glucokinase gene: isolation,

characterization, and identification of two missense mutations linked to early-onset non-insulin-dependent (type 2) diabetes mellitus. Proc Natl Acad Sci 1992; 89:7698-7702.
56. Magnuson MA. Tissue-specific regulation of glucokinase gene expression. J Cell Biol 1992; 48:115-121.
57. Leff SE, Rosenfeld MG, Evans RM. Complex transcriptional units: diversity in gene expression by alternative RNA processing. Annu Rev Biochem 1986; 55:1091-1117.
58. Schibler U, Sierra F. Alternative promoter in developmental gene expression. Annu Rev Genet 1987; 21:237-257.
59. López-Casillas F, Kim K-H. Heterogeneity at the 5' end of rat acylcoenzyme A carboxylase mRNA. Lipogenic conditions enhance synthesis of a unique mRNA in liver. J Biol Chem 1989; 264:7176-7184.
60. Gidh-Jain M, Takeda J, Xu LZ et al. Glucokinase mutations associated with non-insulin-dependent (type 2) diabetes mellitus have decreased enzymatic activity: Implications for structure/function relationships. Proc Natl Acad Sci USA 1993; 90:1932-1936.
61. Hayzer DJ, Iynedjian PB. Alternative splicing of glucokinase mRNA in rat liver. Biochem J 1990; 270:261-263.
62. Magnuson MA. Glucokinase gene structure. Functional implications of molecular genetic studies. Diabetes 1990; 39:523-527.
63. Smith CWJ, Patton JG, Nadal-Ginard B. Alternative splicing in the control of gene expression. Annu Rev Genet 1989; 23:527-577.
64. Tanizawa Y, Matsutani A, Chiu KC et al. Human glucokinase gene: isolation, structural characterization and identification of a microsatellite repeat polymorphism. Mol Endocrinol 1992; 6:1070-1081.
65. Matsutani A, Janssen R, Donis-Keller H et al. A polymorphic $(CA)_n$ repeat element maps the human glucokinase GCK gene to chromosome 7p. Genomics 1992; 12:319-325.
66. O'Rahilly S, Wainscoat JS, Turner RC. Type II (non-insulin dependent) diabetes mellitus: new genetics for old nightmares. Diabetologia 1988; 31:407-414.
67. Froguel P, Vaxillaire M, Sun F et al. Close linkage of glucokinase locus on chromosome 7p to early-onset non-insulin-dependent diabetes mellitus. Nature 1992; 356:162-164.
68. Fajans SS. Scope and heterogeneous nature of MODY. Diabetes Care 1990; 13:49-64.
69. Hattersley AT, Turner RC, Permutt MA et al. Type 2 diabetes is linked to the glucokinase gene in a large pedigree. Lancet 1992; 339:1307-1310.
70. Vionnet N, Stoffel M, Takeda J et al. Nonsense mutation in the glucokinase gene. Nature 1992; 356:721-722.
71. Stoffel M, Patel P, Lo Y-MD et al. Characterization of a missense hexokinase mutation in maturity-onset diabetes of the young (MODY) and mutation screening in late-onset diabetes. Nature Genet 1992; 2:153-156.
72. Kacser H, Burns JA. The molecular basis of dominance. Genetics 1981; 97:639-666.
73. Zouali H, Vaxillaire M, Lesage S et al. Linkage analysis and molecular scan-

ning of glucokinase gene in NIDDM families. Diabetes 1993; 42:1238-1245.
74. Tanizawa Y, Chiu KC, Province MA et al. Two microsatellite repeat polymorphisms flanking opposite ends of the human glucokinase gene: use in haplotype analysis of Welsh Caucasians with type 2 (non-insulin-dependent) diabetes mellitus. Diabetologia 1993; 36:409-413.
75. Hattersley AT, Saker PJ, Cook JT et al. Microsatellite polymorphisms at the glucokinase locus: a population association study in Caucasian type 2 diabetic subjects. Diabet Med 1993; 10:694-698.
76. Elbein SC, Hoffman M, Chiu K et al. Linkage analysis of the glucokinase locus in familial type 2 (non-insulin-dependent) diabetes pedigrees. Diabetologia 1993; 36:141-145.
77. Velho G, Froguel P, Clément K et al. Primary pancreatic beta-cell secretory defect caused by mutations in glucokinase gene in kindreds of maturity onset diabetes of the young. Lancet 1992; 340:444-448.
78. Martin BC, Warram JH, Krolewski AS et al. Role of glucose and insulin resistance in development of type 2 diabetes mellitus: results of a 25-year follow-up study. Lancet 1992; 340:925-929.
79. Forster AC, Symons RH. Self-cleavage of plus and minus RNAs of a virusoid and a structural model for the active sites. Cell 1987; 49:211-220.
80. Haseloff J, Gerlach WL. Simple RNA enzymes with new and highly specific endoribonuclease activities. Nature 1988; 334:585-591.
81. Sakura H, Eto K, Kadowaki H et al. Structure of the human glucokinase gene and identification of a missense mutation in a Japanese patient with early-onset non-insulin-dependent diabetes mellitus. J Clin Endocrinol Metab 1992; 75:1571-1573.

CHAPTER 6

Instantaneous Regulation

An essential point in the regulation of metabolism is how to produce a sufficiently sensitive response to a signal: how can a metabolic pathway respond to an effector (a substrate, an activator or an inhibitor) with high sensitivity? The high half-saturation concentration of hexokinase D in relation to the physiological glucose concentration, and especially its kinetic cooperativity with respect to glucose, allow it to respond sensitively to fluctuations of glucose concentration, and consequently allow the corresponding pathways to adjust their rates. This kinetic cooperativity with respect to glucose has a double interest, both physiological in relation to the instantaneous regulation of glucose phosphorylation in response to the glucose supply, and theoretical, as kinetic cooperativity is a very unusual property in a monomeric enzyme. It requires a kinetic explanation, because the classical "quasi-equilibrium" models of cooperativity cannot account for cooperativity in a monomeric enzyme with only one binding site. Thus in this chapter I discuss the physiological importance of glucose cooperativity and describe the models that have been proposed to explain it. Now that it is possible to produce hexokinase D mutants and to obtain sufficient enzyme for structural studies, there is no doubt that further testing and development of these models is both possible and desirable.

REQUIREMENTS FOR HIGH SENSITIVITY IN THE RESPONSE TO GLUCOSE

For a metabolic reaction to respond adequately to physiological changes of substrate concentration there are some requirements that must be satisfied. First of all the half-saturation concentration must be high enough for the enzyme not to be already saturated at the low end of the range of substrate concentrations, so that the rate can increase when the substrate concentration increases. In other words the half-saturation concentration in vivo ought to be higher than the lowest physiological concentration.

However, satisfying this very important requirement may not be enough, as the system needs not merely to respond to fluctuations in the substrate concentration, but to do so with high sensitivity, and this is not possible with ordinary Michaelis-Menten kinetics. For an enzyme obeying the Michaelis-Menten equation a ten-fold increase in rate requires at least a ten-fold increase in substrate concentration, and much more if the concentrations are not very small; to cover the middle 10-90% range of possible rates requires an 81-fold increase in substrate concentration—almost two orders of magnitude. As in most cases the physiological variation in substrate concentration is much smaller than this, the corresponding variation in rate must be very small for a Michaelis-Menten enzyme. Cooperativity in the response of the enzyme to the effector, substrate in this case, is a mechanism that allows a more sensitive response.

As we saw in Chapter 3, hexokinase D satisfies both requirements. It not only has a half-saturation concentration for glucose higher than the level of circulating glucose before a meal, but it also displays kinetic cooperativity with respect to glucose. These properties allow the physiological fluctuations of glucose concentration expected in the hepatocyte, especially in animals that feed intermittently, to be translated automatically into variations in hexokinase D activity; thus the utilization of glucose is automatically adjusted in relation to its supply in the diet. The other mammalian hexokinases behave quite differently. Their low K_m values and lack of cooperativity mean that they are saturated at low glucose concentrations and their activities do not increase with glucose concentration (Fig. 6.1). In addition, their high sensitivity to glucose 6-phosphate means that they may be largely inhibited at the glucose 6-phosphate concentration believed to exist in the hepatocyte, which is in the range 0.1-0.3 mM.[1]

A Hill coefficient of 1.6 may seem too small to have a physiological meaning, but it means that a 16-fold (instead of 81-fold) increase in substrate concentration is sufficient to increase the rate from 10 to 90% of the maximum: in other words the ratio of substrate concentrations needed to span the middle 10-90% range of rates is 5-fold smaller than it is for an enzyme obeying Michaelis-Menten kinetics. This is by no means trivial as an explanation of the physiological advantage of the cooperative behavior of hexokinase D.

INTERPRETATION OF KINETIC COOPERATIVITY

The idea that has dominated thinking about cooperativity is the allosteric model suggested by Monod and coworkers,[2,3] which was modified and extended by Koshland and coworkers.[4] Later reviews are references 5 and 6. The classic models emphasize the importance of interactions between subunits in the control of enzyme activity, so that sigmoid kinetics is interpreted in terms of cooperative binding of ligands to interacting subunits of the enzyme, with maintenance of equilibrium between the different enzyme species. Much of the data in the literature are consistent with these theories, and because they require multiple subunits it has become widely believed that sigmoid kinetics must be restricted to oligomeric enzymes.

Hexokinase D, however, is a monomeric enzyme, as we saw in Chapter 3, and many readers may feel puzzled about its cooperativity with respect to glucose, or even irritated at the use of the word cooperativity. They may feel, as mentioned above, that any idea of cooperation requires the existence of multiple binding sites. It is for this reason that I have been always talking in terms of *kinetic* cooperativity, to make it explicit that this behavior of hexokinase D is not the consequence of ligand binding at equilibrium.

Despite the influence of the classic equilibrium models of cooperativity, it has been known for about 30 years that monomeric enzymes can in principle show kinetic behavior similar in appearance to the binding of ligands to oligomeric proteins in which there are cooperative interactions between multiple binding sites.[7] Non-equilibrium models do not require subunit interactions, and may therefore be applied to monomeric enzymes,

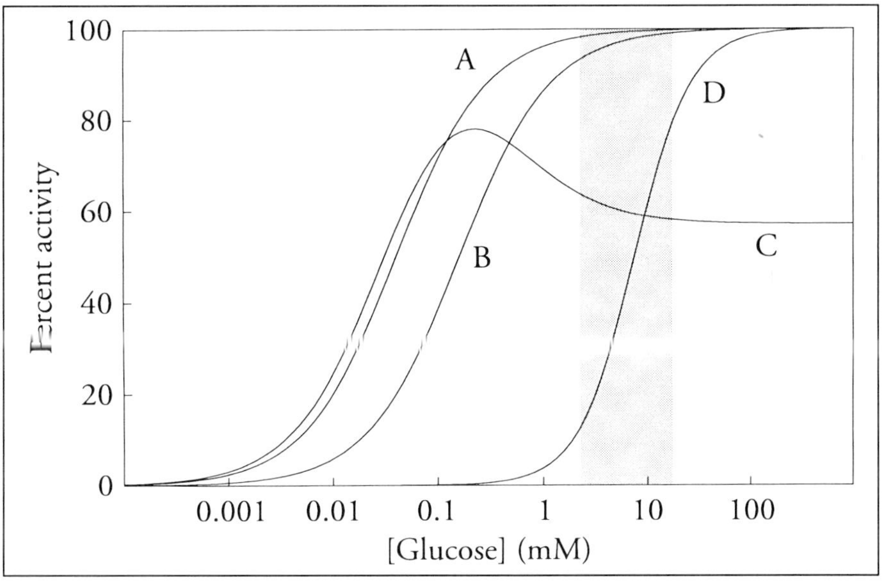

Fig. 6.1. Kinetic characteristics of hexokinase isoenzymes in the physiological range of glucose concentrations. The shaded area represents the ordinary range of glucose concentrations expected in the hepatocyte; within this range hexokinases A and B are essentially saturated, so that their activity is virtually independent of the glucose concentration, and any variation is to some degree counteracted by variation in the opposite direction of the activity of hexokinase C. Only hexokinase D shows a strong dependence of activity on glucose concentration in this range. The curves were calculated assuming that hexokinases A and B obey Michaelis-Menten kinetics with K_m values of 0.04 and 0.16 mM respectively, that hexokinase C displays substrate inhibition and hexokinase D shows cooperativity with half saturation at 7.5 mM and a Hill coefficient of 1.6. In a semilogarithmic plot any saturation curve looks S-shaped, the degree of cooperativity being evident in the steepness of the curve at the midpoint; thus the curves for hexokinase A and B are S-shaped (but less steep than that for hexokinase D), despite the lack of cooperativity with respect to glucose.

as reviewed elsewhere.[8] In such models kinetic cooperativity results from the existence in the steady state of different enzyme states that are not in equilibrium with each other. In this field the theoretical models preceded the experimental examples by more than 10 years; hexokinase D was the first reported case of a monomeric enzyme with sigmoidal kinetics, and by that time the dogma was so well established that it has not been easy to eradicate it. The scarcity of examples may reflect Ricard's view that cooperativity in an enzyme with only a single binding site can only be achieved at the expense of catalytic efficiency,[9] but it may also be the consequence of lack of recognition of kinetic cooperativity when it has been observed. For further discussion and examples see the review of Cornish-Bowden and Cárdenas.[8]

As we'll describe below, several kinetic models of cooperativity have been applied to hexokinase D, but with the present evidence it is difficult to establish which is the right one, a fact that is well illustrated by the papers of Pettersson.[10,11] My own ideas have evolved with time and in the light of new experimental evidence. We were initially attracted by the random mechanism, but subsequently we included a conformational change to better explain the effect of *N*-acetylglucosamine,[12] finally arriving at a slow transition model with ordered binding of substrates.[13,14] Now that it has become possible to have enough pure enzyme to do experiments that were not feasible when the models were proposed, it would be useful for some definitive experiments to be done.

HEXOKINASE D: A MONOMERIC ENZYME WITH ONLY ONE BINDING SITE FOR GLUCOSE

Is hexokinase D a monomer in all conditions?

For a monomeric enzyme like hexokinase D it is crucial not only to demonstrate that the enzyme is a monomer but that it remains a monomer under the conditions used for the kinetic measurements and, most important, that there is no dependence of molecular mass on the concentration of the substrate showing cooperativity. This requirement has apparently not been satisfied with many enzymes, but it has been done with hexokinase D, using different methods. For example dimethyl suberimidate, a bifunctional cross-linking reagent,[15] was used to determine whether the reacting enzyme could form dimers.[16] Incubations were set up under assay conditions, the protein was "frozen" by the addition of the bifunctional reagent and the cross-linked enzyme was then subjected to polyacrylamide electrophoresis in the presence of sodium dodecylsulfate. The authors did not specify which conditions they tested, but they stated that in all of them a single protein band migrated on these gels corresponding to a molecular mass of 52 kDa, and there was no suggestion of any higher molecular mass cross-linked material.

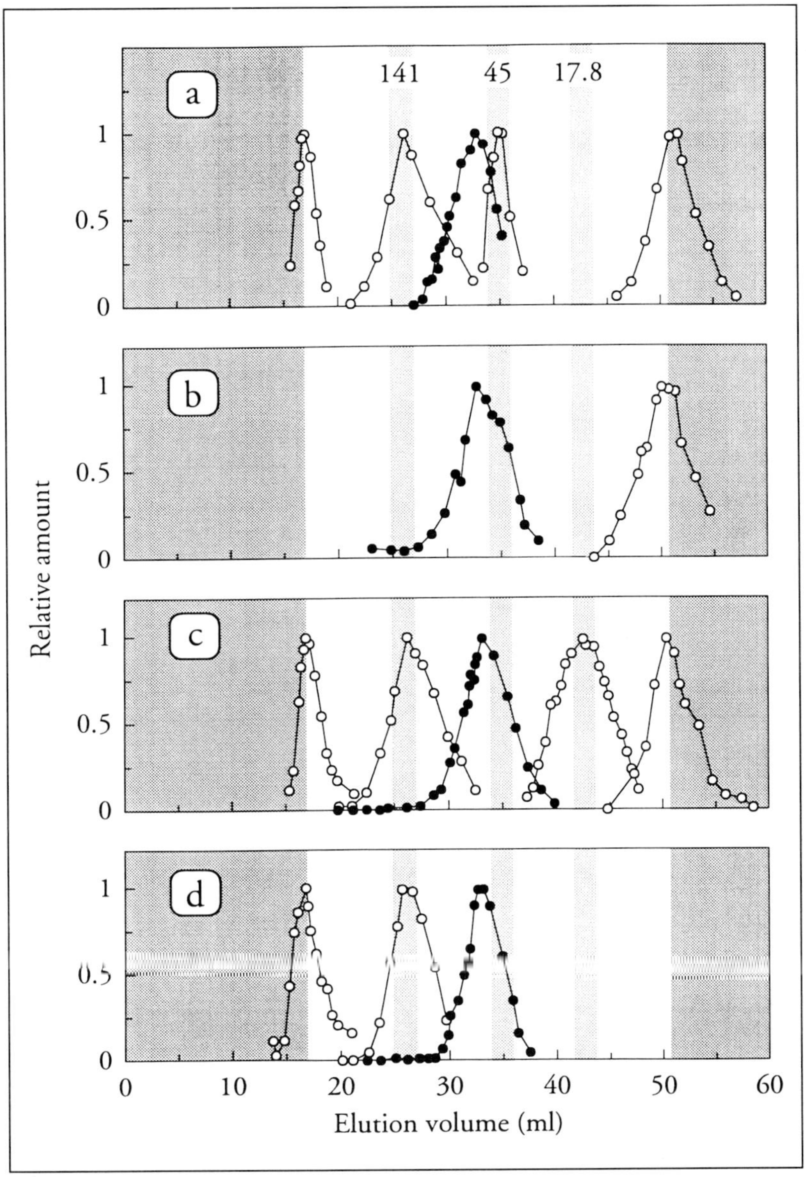

Fig. 6.2. Elution profiles for hexokinase D in the presence of substrates. Partially purified hexokinase D (●) was eluted from Sephadex G-200, with no added substrates (a), or in the presence of (b) 5 mM ATP, (c) 5 mM ATP and 1 mM glucose, (d) 5 mM ATP and 100 mM glucose. The shading and additional profiles (○) indicate the markers used: blue dextran, yeast alcohol dehydrogenase (141 kDa), ovalbumin (45 kDa), myoglobin (17.8 kDa) and [^{14}C]leucine (131 Da). Adapted from Figure 5 of Arch Biol Med Exp 1979; 12:571-580.

As this was a crucial point, we reinvestigated the problem by gel filtration under reacting conditions, and again found that rat hexokinase D is a monomeric protein that does not polymerize under conditions similar to those of the assay.[17] Thus, in agreement with the results with suberimidate, gel filtration at 30° C in the presence of glucose and MgATP, singly and in combination, at both low (1 mM) and high (100 mM) glucose concentrations, showed only one symmetrical peak of activity (Fig. 6.2). The apparent molecular mass of 54 kDa estimated under these conditions agreed well with the values reported for hexokinase D in the presence of denaturing agents,[16,18,19] and with the value calculated from the cDNA sequence.[20,21]

Likewise, the presence of the products of the reaction, glucose 6-phosphate and MgADP, did not change the distribution coefficient of hexokinase D. The competitive inhibitor *N*-acetylglucosamine, which does change the kinetic behavior of hexokinase D, as we'll see shortly, does not modify the monomeric nature of the enzyme either.[17] Nor does glycerol,[14] though it alters the kinetic behavior of hexokinase D by decreasing the cooperativity (see below).

The persistence of the monomeric form of hexokinase D in all these experiments excludes the possibility that the cooperativity results either from the interaction of subunits in an oligomer or from an association-dissociation equilibrium between different states of the enzyme with different properties for the different molecular sizes. Moreover, the suppression of the cooperativity by glucose analogs or by glycerol cannot be explained by a change in the molecular mass.

Only one binding site for glucose?

A important remaining question, however, was whether despite its monomeric structure hexokinase D might have more than one active site per molecule. This possibility was excluded by affinity labelling with the true affinity label *N*-(*N*-bromo[1-^{14}C]acetyl)-6-aminohexanoylglucosamine.[22] Although two molecules of this reagent can be incorporated per molecule of hexokinase D, only one binds at the active site, because when the experiments were done in the presence of saturating *N*-acetylglucosamine (70 mM), only 1 molecule of reagent was incorporated per molecule of enzyme, without loss of the catalytic activity. (Both glucose and *N*-acetylglucosamine protect competitively against inactivation). According to the authors the binding to the second site appears to be very unspecific.[22]

A point that could be raised about this experiment is that, even if there is only one active site, binding of glucose at a second site that interacted with the first could generate cooperativity. An obvious experiment to answer this question would be to study the kinetic behavior of the enzyme after alkylation of the non-active site. Unfortunately this appears not to have been done, probably because the aim of the paper was to compare the active sites of

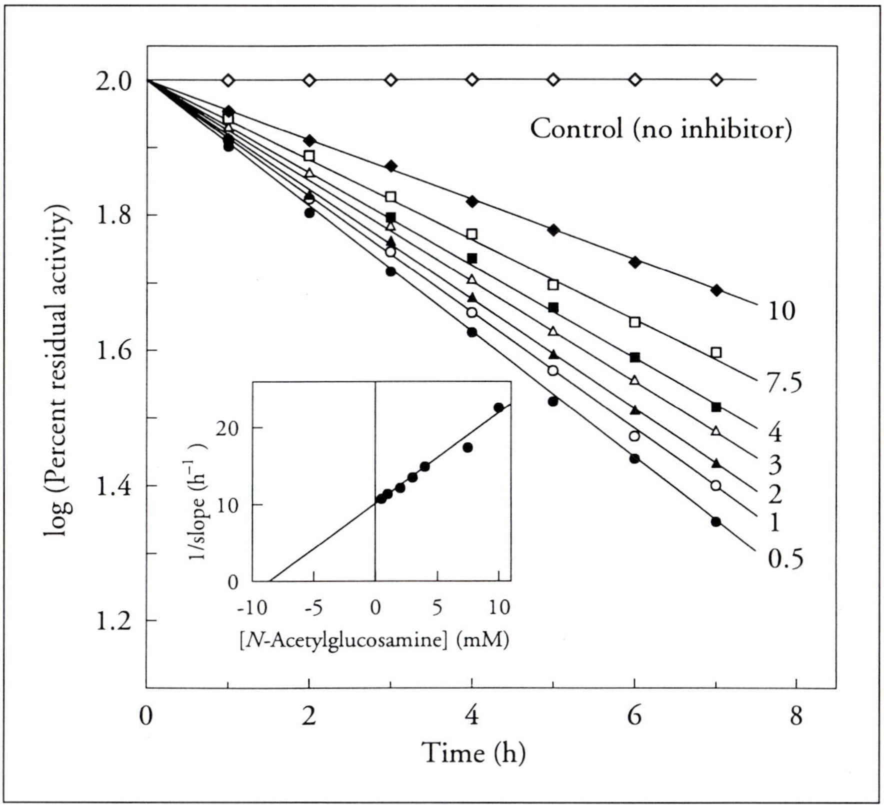

Fig. 6.3. Protection of hexokinase D against inactivation. Hexokinase D was inactivated by an affinity label, N-(N-*bromoacetyl)-6-aminohexanoylglucosamine in the presence of various concentrations of* N-*acetylglucosamine as follows: 0.5 mM (●); 1 mM (○), 2 mM (▲); 3 mM (Δ); 4 mM (■); 7.5 mM (□); 10 mM (◆). The control (◊) was done in the absence of affinity label. The inset shows a replot of slopes against the* N-*acetylglucosamine concentration. Reproduced with permission from Eur J Biochem 1979; 49:299-308.*

hexokinase D and hexokinase B. The hypothesis of two interacting binding sites for glucose can be discarded anyway because the protection experiments using glucose or *N*-acetylglucosamine gave a linear dependence on the sugar concentration (Fig. 6.3).[22] In fact, corresponding linearity has always been observed in experiments of protection against inactivation, which are an indication of glucose binding.[23] Furthermore, in measurements of intrinsic fluorescence, titration of hexokinase D with glucose gives a hyperbolic curve, confirming that the binding of glucose in equilibrium conditions is not cooperative (Fig. 6.4).[14,24] The deviation from the line at the lowest concentration is probably due to the fact discussed below that the enhancement of the intrinsic fluorescence of hexokinase D by addition of glucose requires some time, which increases as the glucose concentration decreases, so that this value could have been underestimated.

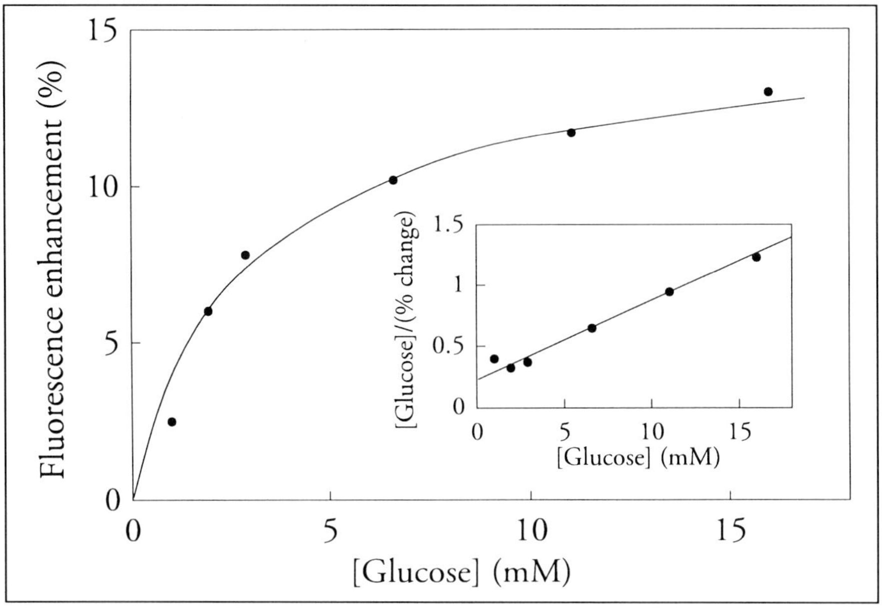

Fig. 6.4. Effect of glucose on the fluorescence of hexokinase D. Fluorescence emission was measured at 326 nm, with excitation at 284 nm, in the presence of different concentrations of glucose. The secondary plot shown in the inset indicates a dissociation constant of 3.2 mM for glucose. Reproduced with permission from Arch Biol Med Exper 1985; 18:273-284.

It can therefore be concluded that despite the kinetic cooperativity with respect to glucose, hexokinase D is a monomeric enzyme even under reaction conditions, and has no interacting glucose binding sites. As the binding of glucose at equilibrium appears to be hyperbolic, consistent with this absence of interactions between sites, the cooperativity of glucose phosphorylation must be kinetic in origin. However, before introducing the models that could explain this kinetic cooperativity I shall describe the conditions in which it is decreased or even eliminated, as knowledge of these conditions is important for considering the models.

CONDITIONS THAT DECREASE THE POSITIVE COOPERATIVITY OF HEXOKINASE D WITH RESPECT TO GLUCOSE

SUPPRESSION OF GLUCOSE COOPERATIVITY BY COMPETITIVE INHIBITORS

If the cooperativity were derived from a conformational transition induced or modified by glucose, an alternative sugar substrate that shows cooperativity by itself, such as mannose, should be able to "collaborate" with glucose in favoring the existence of a single conformation and thus decreasing the cooperativity. With this in mind we studied the saturation function of hexokinase D with glucose as substrate in the presence of other substrates. These alternative sugar substrates, mannose, fructose and 2-deoxyglucose, together with the analog *N*-acetylglucosamine, which is not phosphorylated by hexokinase D, all behave as competitive inhibitors with respect to glucose,[12,13,25-28] which is not surprising. What is surprising, however, is that despite their different kinetic behavior when used as substrates, mannose, fructose and 2-deoxyglucose are all able to suppress cooperativity with glucose, albeit with different efficiencies as shown in Table 6.1.[13,28] Although hexokinase D has very low affinity for fructose and shows no cooperativity with it as substrate, as mentioned in Chapters 2 and 3, fructose was the most efficient in suppressing the glucose cooperativity. On the other hand, *N*-acetylglucosamine was the least efficient, despite being the best competitive inhibitor of glucose phosphorylation (K_i = 0.14 mM). The suppression of cooperativity by the inhibitor is a specific effect, not an unspecific result of an increase in the concentration of alcohol groups or of the viscosity produced merely by the increased sugar concentration in the assay medium. This follows not only from the different concentrations of inhibitor required in each case, but also from the fact that galactose, which is neither a substrate nor an inhibitor, had no effect on the cooperativity even at a concentration as high as 320 mM.[13]

Long-chain acyl-coenzymes A, which are competitive inhibitors of hexokinase with respect to both glucose and MgATP, have no effect on the positive cooperativity with respect to glucose, as discussed in Chapter 7. The inhibition by *N*-acetylglucosamine and mannose at low glucose concentration deserves some comment, however, especially as *N*-acetylglucosamine is widely used as a competitive inhibitor of hexokinase D.

Table 6.1. Efficiency of glucose analogs for suppressing the cooperativity of hexokinase D with respect to glucose[1]

Analog	[A] [2] (mM)	K_i [3] (mM)	$K_{0.5}$ [4] (mM)	[A]/K_i	[A]/$K_{0.5}$
Fructose	250	107	325	≈2	≈1
Mannose	25	3.1	8	8	3
2-Deoxyglucose	200	n.d.	22	–	10
N-Acetylglucosamine[5]	12	0.14	—	85	—

1 Data of Cárdenas and coworkers.[11,26]
2 Analog concentration necessary for obtaining a Hill coefficient of 1.0 with respect to glucose.
3 Competitive inhibition constant obtained from a Dixon plot when the analog was treated as a competitive inhibitor with respect to glucose.
4 Half-saturation concentration when the analog was used as substrate.
5 *N*-Acetylglucosamine is not a substrate.

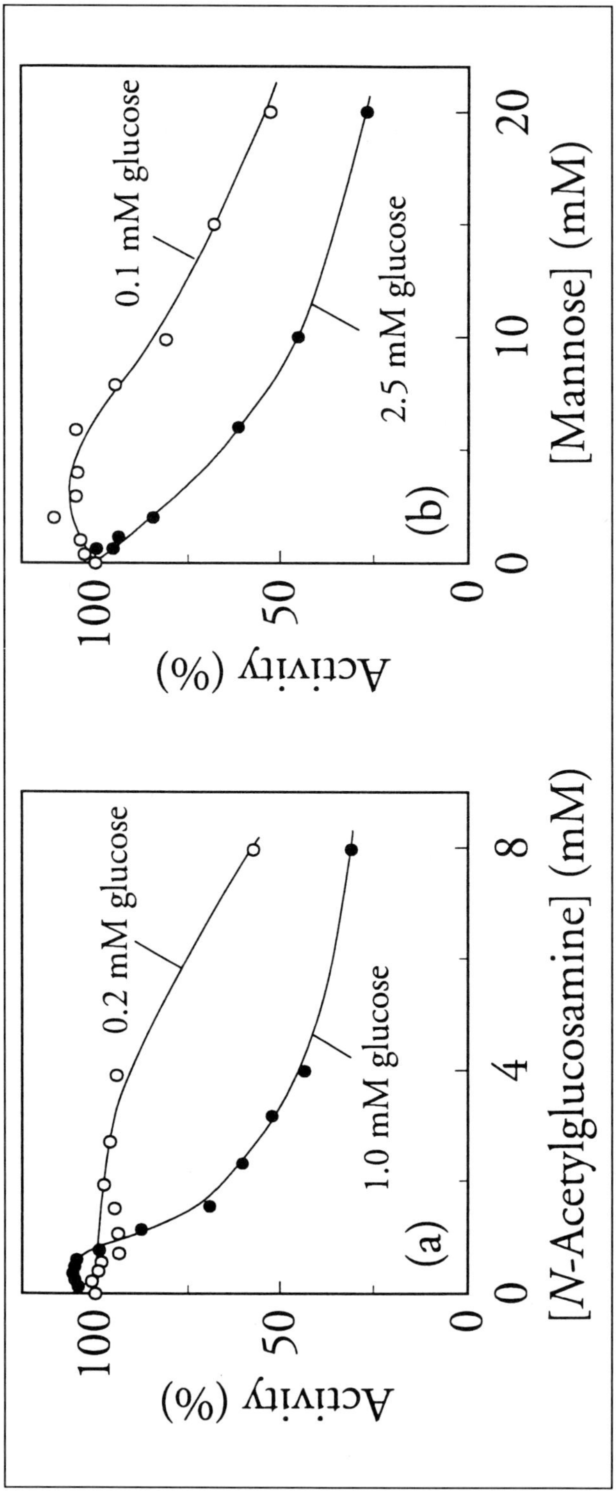

Fig. 6.5. Cooperativity of inhibition of hexokinase D by (a) N-acetylglucosamine and (b) mannose at low glucose concentrations. At low concentrations of glucose the inhibition by both inhibitors increases with the glucose concentration (the opposite of the expected behavior for a competitive inhibitor), and at very low glucose concentrations there is virtually no effect of either inhibitor at concentrations where strong inhibition would be expected if extrapolation of the behavior at high glucose concentrations were valid. Reproduced with permission from Eur J Biochem 1984; 145:163-171.

Inhibition by *N*-acetylglucosamine

With this inhibitor Dixon plots are essentially linear if concentrations of glucose higher than 4 mM are used, and in these conditions a K_i value of about 0.14 mM at pH 7.5 or 8.0 can be estimated.[13] This value is lower than those of 0.38-0.5 mM reported in earlier studies,[25-27] probably for the same reasons as those discussed in Chapter 3 in relation to the half-saturation concentration. On the other hand at low glucose concentrations (<2 mM) Dixon plots give curves with a downward curvature that increases as the glucose concentration decreases, mimicking partial competitive inhibition. The characteristics of the inhibition suggest that *N*-acetylglucosamine interacts with different affinities with more than one enzyme form, and that the relative proportions of these forms vary with the glucose concentration. Thus at glucose concentrations below about 1-2 mM the inhibition by *N*-acetylglucosamine over a wide range of concentrations around the inhibition constant increases as the glucose concentration increases, not at all the behavior expected for a competitive inhibitor. The decrease in the inhibitory effect of *N*-acetylglucosamine as the glucose concentration decreases is illustrated in Figure 6.5a: an *N*-acetylglucosamine concentration as high as 3 mM (about 20 K_i) produces almost no inhibition at 0.2 mM glucose, but about 50% inhibition at 1 mM glucose. At this last concentration of glucose a small activation (about 10%) is observed at low *N*-acetylglucosamine

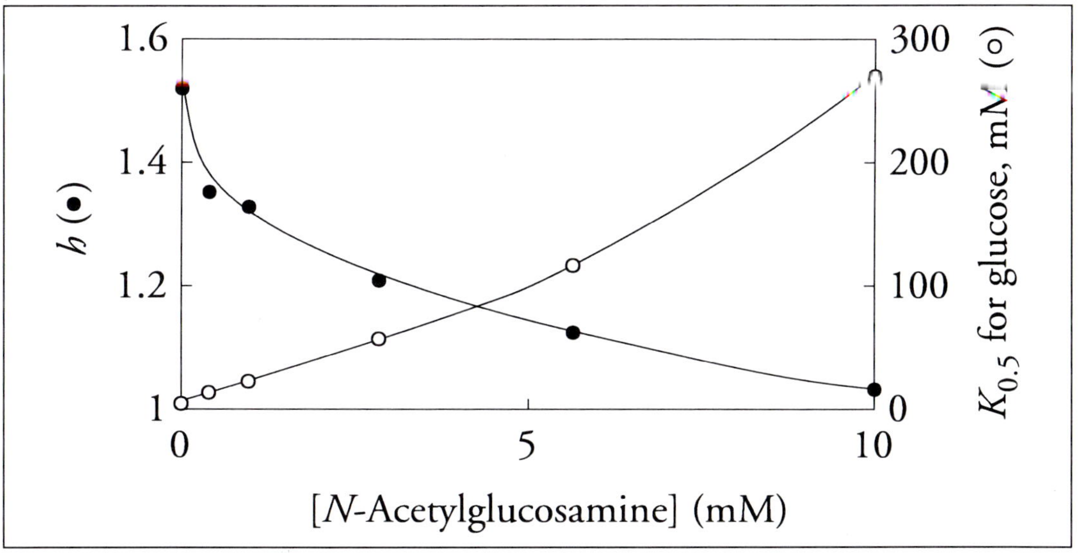

Fig. 6.6. Effect of N-acetylglucosamine *on the cooperativity of hexokinase D with respect to glucose. As the* N-acetylglucosamine *concentration increases the Hill coefficient approaches a value of 1 (no cooperativity) and there is a large increase in the half-saturation concentration for glucose. Adapted from Arch Biol Med Exp 1979; 12:571-580.*

concentrations. There is practically no effect of *N*-acetylglucosamine in experiments at low glucose concentration (0.2 mM) unless its concentration increases above 4 mM, i.e. over 25K_i. This phenomenon can be described as cooperativity of *N*-acetylglucosamine inhibition.

The lack of inhibition at low glucose concentration is not an artifact derived from transformation of *N*-acetylglucosamine during the assay, because if aliquots are withdrawn after several minutes of reaction and used in assays with higher glucose concentrations the expected inhibition is obtained. When the saturation function for glucose was

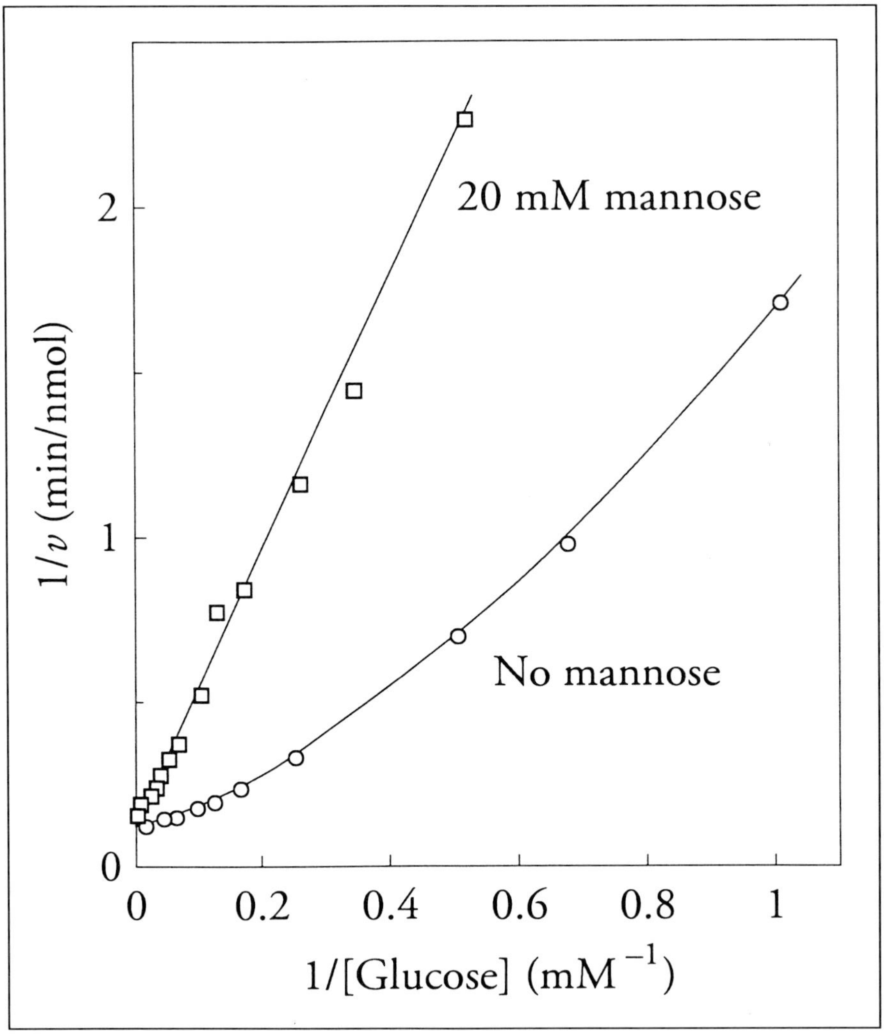

Fig. 6.7. Effect of mannose on the kinetics of hexokinase D with respect to glucose. 20 mM mannose eliminates the cooperativity with respect to glucose, giving a straight line in the double-reciprocal plot. Redrawn from Eur J Biochem 1984; 145:163-171.

studied in the presence of increasing concentrations of *N*-acetylglucosamine the Hill coefficient decreased progressively (Fig. 6.6), with a value of 1.0 expected at about $86K_i$ (Table 6.1).

Inhibition by mannose

As with *N*-acetylglucosamine, increasing the mannose concentration produces a progressive decrease in the Hill coefficient for glucose and an increase in the half-saturation concentration, mannose being considerably more effective than *N*-acetylglucosamine for suppressing the cooperativity (Table 6.1). At 20 mM mannose there was no cooperativity (Fig. 6.7), as a straight line was obtained in the double-reciprocal plot. Again like *N*-acetylglucosamine, mannose activates glucose phosphorylation at very low glucose concentrations (Fig. 6.5b),[13] to a decreasing extent as the glucose concentration increases. This cooperativity in the inhibition by mannose again implied that to calculate the inhibition constant from a Dixon plot required use of data at glucose concentrations greater than the half-saturation concentration, in which case reasonable straight lines could be obtained, providing a K_i of 3.1 mM, lower than values in the range 14-16 mM reported in earlier work,[26,27] probably for the same reasons discussed in Chapter 3 in relation to $K_{0.5}$.

The cooperativity of the inhibition means that these K_i values are not entirely satisfactory, as they are limiting values for glucose concentrations that allow the existence of hexokinase D in a single conformation, as discussed below. An important point is that any of the values in the range 0.14-0.5 mM reported by different groups for K_i for *N*-acetylglucosamine is much lower than the dissociation constant of 4.6 mM obtained in the protection experiments illustrated in Figure 6.3,[illegible] the corresponding K_i value of the same authors being 0.26 mM. Although it is readily recognized that kinetically determined K_i values are not the same as true dissociation constants, at the time the authors were unable to offer any reasonable explanation for these differences. I think that the discrepancy can now be explained on the basis of the differences in the conformation of hexokinase D in the presence and absence of glucose, as will be further discussed below.

Inhibition by 2-deoxyglucose and fructose

Like *N*-acetylglucosamine and mannose, 2-deoxyglucose and fructose suppress cooperativity, but experiments at very low glucose concentrations similar to those just described are not possible because of difficulties in estimating the rate with sufficient accuracy (see the section on assays in Chapter 3). The effectiveness of fructose in suppressing cooperativity will be discussed further in relation to the models.

Effect of Glycerol

As mentioned already hexokinase D is an unstable enzyme. As glycerol was known to be effective for preserving enzyme stability

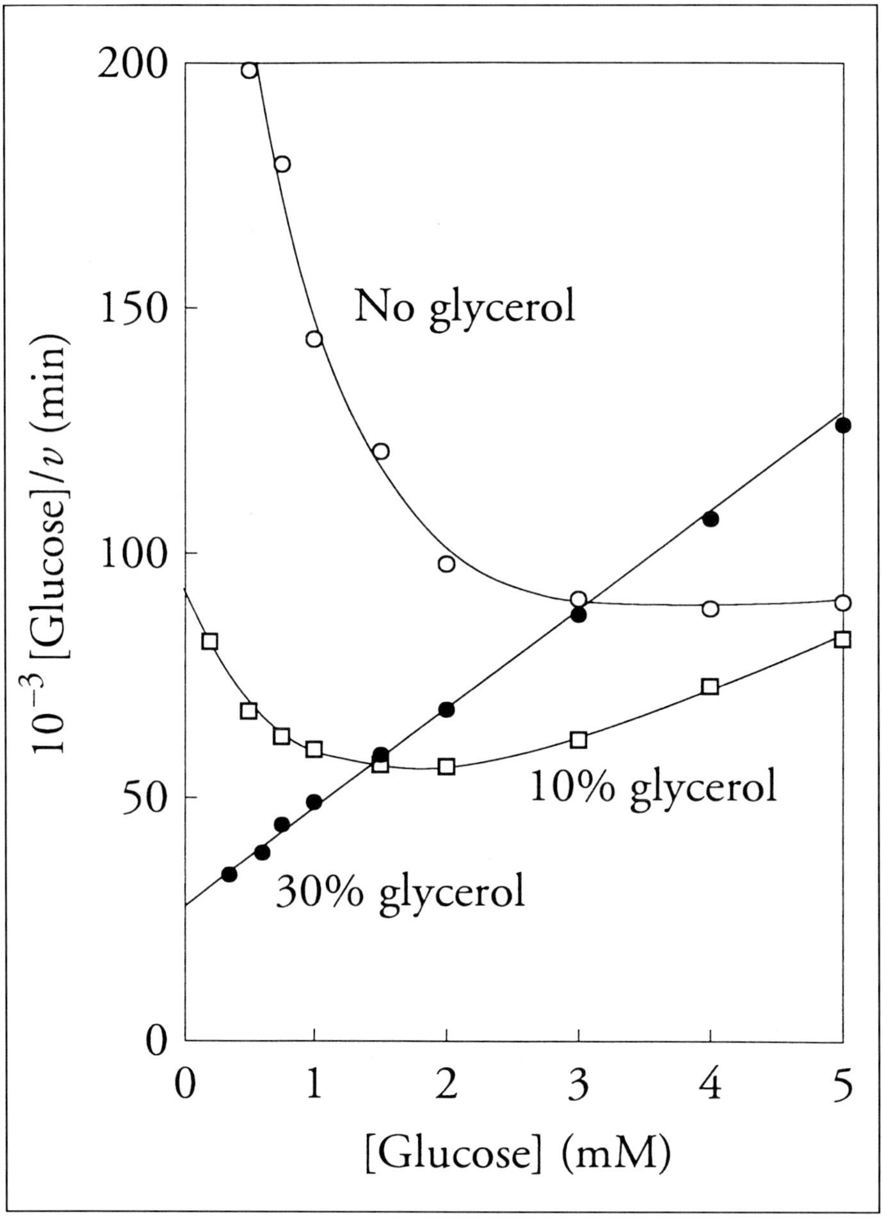

Fig. 6.8. Effect of glycerol on the cooperativity of hexokinase D with respect to glucose. Addition of glycerol decreases and, at concentrations above 20% (v/v), eliminates the cooperativity. Redrawn from Arch. Biochem. Biophys. 1985; 237:328-334. The original figure included data at 5% (intermediate between 0 and 10%) and 20% (similar to 30%) glycerol, which are here omitted in the interests of clarity.

during storage,[29,30] it was introduced into the purification in the hope of improving its stability, with great success.[16] In fact 30% glycerol used routinely for storing the enzyme strongly protects it from sulfhydryl group oxidation and inactivation.[31] This stabilizing effect suggested immediately that it might also modify its kinetic behavior by affecting the conformation of hexokinase D, especially as it was being used in high concentrations (10-30% v/v).

When the experiments were finally done,[14,32] the degree of cooperativity was found to be dependent on the presence of glycerol in the assay mixture, decreasing progressively until it disappeared completely at concentrations of about 20% (v/v) glycerol (Fig. 6.8).[32] In 20% and 30% (v/v) glycerol the slope of the Hill plot was unity, i.e. there was no detectable cooperativity. Notice in Figure 6.8 that glycerol activated at low glucose concentrations but inhibited at high glucose concentration: at 30% glycerol there is a decrease of maximal rate and of the half-saturation concentrations of glucose and MgATP by about 50-60%, the reduction of the $K_{0.5}$ value explaining the activation at low glucose concentration. The Michaelis-Menten type kinetics with respect to MgATP were not modified in the presence of glycerol at a concentration as high as 30%.[32]

All of these effects of glycerol were fully reversible, so that glycerol (or any possible impurity, e.g. aldehydes) does not cause chemical modification of the enzyme. The reversibility by dilution was very important, as satisfactory storage conditions of hexokinase D required 30% glycerol.

Gekko and Timasheff[33,34] have investigated the nature of the interactions between glycerol and proteins. They have found that in aqueous solutions glycerol causes proteins to become preferentially hydrated and is itself excluded from the domain of the protein. Addition of glycerol therefore has the thermodynamically unfavorable effect of raising the chemical potential of the system. This effect can be regarded as distributed over the entire protein solvent-interface and so a reduction of this interface will be thermodynamically favorable. Glycerol should then tend to favor the more folded or compact state.

The effect of glycerol is probably to freeze the enzyme in a conformation with high affinity for glucose, preferentially stabilizing it, as will be further discussed below in relation to the kinetic models proposed to explain kinetic cooperativity.

Solvent isotope effects on the hexokinase D reaction

Effect of ^2H_2O (D_2O)

^2H_2O can sometimes have drastic effects on protein structural equilibria, and for this reason Cornish-Bowden studied the kinetics of the hexokinase D reaction in ^2H_2O.[35] In contrast to the positive cooperativity with respect to glucose observed in ^1H_2O, there is negative cooperativity in ^2H_2O; at high glucose concentration there is a normal solvent isotope effect (reaction faster in ^1H_2O than in ^2H_2O), whereas at low glucose concentrations and high concentrations of MgATP there is an inverse solvent isotope effect (reaction faster in ^2H_2O). The half-saturation points for both glucose and MgATP are lower in ^2H_2O than in ^1H_2O, decreasing from 5 to 3.5 mM in the case of glucose. The reaction followed Michaelis-Menten kinetics with respect to MgATP at all glucose concentrations in ^2H_2O, just as in ^1H_2O.

The isotope effects will be further discussed below in relation to the models for explaining cooperativity.

MODELS FOR THE KINETIC COOPERATIVITY OF HEXOKINASE D WITH GLUCOSE

Several models can explain kinetic cooperativity in a monomeric enzyme with a single active site, and in principle any of them could be applied to hexokinase D. However, as we shall see shortly, some give a better explanation of the data as a whole.

Random order of substrate binding

In principle any mechanism in which the binding of substrate can occur in two or more pathways in parallel will generate a rate equation containing higher-order terms in the concentration of that substrate. An early example of this is provided by the mechanism for a reaction of two substrates that can bind to the enzyme in either order,

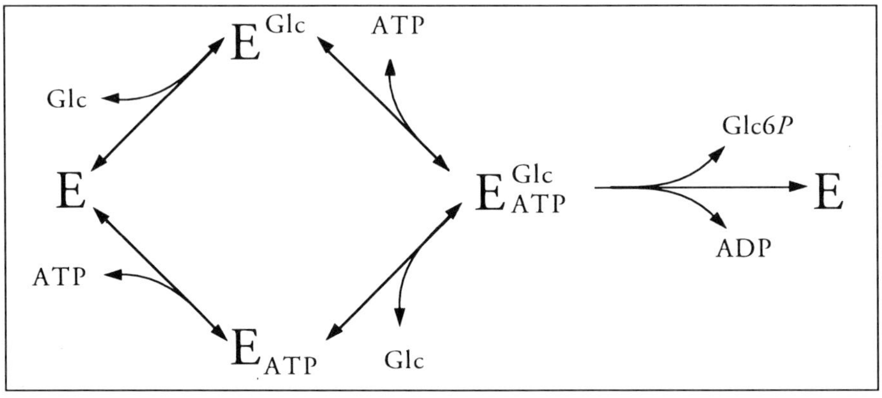

Fig. 6.9. Random-order mechanism for hexokinase D. Cooperativity could in principle arise from a mechanism in which the substrates glucose and MgATP could bind to the enzyme in either order, provided that the ternary complex is capable of reacting fast enough to prevent substrate binding to approach equilibrium. The model predicts that the glucose cooperativity should depend on the MgATP concentration, as observed, but it also predicts deviations from Michaelis-Menten kinetics with respect to MgATP, which are not observed. In principle there may be similar complexities in the order of product release, but these are not shown in the scheme as nearly all studies of the enzyme have been concerned with substrate binding more than with product release.

(Fig. 6.9) in which each substrate can bind to two different enzyme species, the free enzyme and the binary complex formed by the enzyme with the other substrate. In the case of hexokinase D, glucose could bind to the free enzyme or to the binary complex enzyme-MgATP. The rate equation for this mechanism was worked out by Wong and Hanes in 1962,[36] and is discussed in detail in the well known treatise of Dixon and Webb;[37] it is complicated, and contains fourteen different coefficients that represent different combinations of rate constants.

If the concentration of one substrate, e.g. MgATP, is held constant, the dependence of the rate v on the concentration of the other, glucose, takes the following form:

$$v = \frac{N_1[\mathrm{Glc}] + N_2[\mathrm{Glc}]^2}{D_0 + D_1[\mathrm{Glc}] + D_2[\mathrm{Glc}]^2} \tag{1}$$

where the coefficients N_1, N_2, D_0, D_1, and D_2 are linear or quadratic functions of the concentration of MgATP. It does not follow automatically from the presence of higher-order terms in this equation that any particular kind of deviation from Michaelis-Menten kinetics, such as positive cooperativity, is possible, because if both pathways of formation of the ternary enzyme-glucose-MgATP complex were equivalent or if the free enzyme and the three enzyme-substrate complexes were in rapid equilibrium it would be impossible to detect any deviation from Michaelis-Menten behavior.

Ferdinand studied the random-order mechanism in some detail,[7] with a view to establishing the conditions that would lead to cooperativity or substrate inhibition, and showed that in two-substrate reactions sigmoid kinetics may result if the substrates bind to the enzyme at random under non-equilibrium conditions, and if one of the two resultant pathways is kinetically preferred, i.e. it is faster. In terms of the coefficients of Equation (1) this is equivalent to saying that N_1D_0 should be bigger than N_2D_2 and N_1D_2 bigger than N_2D_1. He showed that this model could account well for the regulatory properties of phosphofructokinase, and so as long ago as 1966 it was clear that one did not need to postulate interactions between subunits to explain cooperativity.

This model predicts that the possibility of cooperativity with respect to one substrate depends on the concentration of the other,[7] and in hexokinase D, the cooperativity with respect to glucose does indeed depend on the MgATP concentration (Fig. 3.6). Because of the complexity of the definitions of the coefficients in the full rate equation it is not possible to reduce the condition for cooperativity to a simple form in which the required relationship between the rate constants is obvious. One point is clear, however: cooperativity cannot occur unless the conversion of the ternary complex to products can occur fast enough to prevent substrate binding from reaching equilibrium. If this conversion is fast enough, the proportions of the reaction flux proceeding through the two binary complexes are not constant but vary with the substrate concentration.

The cooperativity would result from binding of glucose to two non-equivalent species, the free enzyme and the binary enzyme-MgATP complex. At high concentrations of MgATP the enzyme would exist mainly as this enzyme-MgATP complex, and the route of reaction at low glucose concentration would be mainly E → E-MgATP → E-MgATP-glucose → products. As the concentration of glucose increased, the other pathway (E → E-glucose → E-MgATP-glucose → products) would become more important and because it is kinetically preferred (faster) this would cause the rate of the reaction to increase more than predicted by the Michaelis-Menten equation, i.e. the rate dependence would be sigmoidal.

On the other hand, at sufficiently low concentrations of MgATP, around $0.2K_m$ or lower, the concentration of the binary enzyme-MgATP complex would be negligible, and so even at low concentrations glucose would bind to the free enzyme, the only form of enzyme available. In this way only the preferred route would operate, which would be seen as Michaelis-Menten kinetics with respect to glucose.

We were initially very attracted by this model as it explained cooperativity in a monomeric enzyme in a simple way, and it also accounted for the dependence of the degree of cooperativity on the MgATP concentration. In addition it appeared plausible in view of indirect evidence from studies of the kinetics of hexokinase D with 2-deoxyglucose as substrate,[38] and of chemical modification of the enzyme and the protection by substrates.[23] At the time there was no experimental information on the order of binding of substrates to hexokinase D, and a random mechanism with a preferred pathway through the binary enzyme-glucose complex accorded well with the type of mechanism widely believed to apply to hexokinases in general.[39] Moreover, the observation that Triton X-100 and urea do not modify the cooperativity of hexokinase D was easily explained in terms of the model. The fact that photooxidation and other treatments did not modify the cooperativity supported this, and implied that the phenomenon of cooperativity was inherent in the catalytic mechanism. Hexokinase D appeared to differ in this respect from wheat hexokinase L_I, for which the negative cooperativity with respect to glucose disappears in the presence of even very small urea concentrations.[55] Simulation studies showed that the model could also account for the suppression of cooperativity by glucose analogs (M.L.C., unpublished results).

To explain the lack of kinetic equivalence between the two pathways at a molecular level we postulated a slow glucose-induced conformational change of the enzyme that occurred at different rates depending on whether glucose bound to the free enzyme or to the enzyme-MgATP complex,[12] because it is difficult to explain in molecular terms a difference in rate between two pathways only on the basis of rate constants for binding of substrates to the free enzyme or to the binary complexes. In general binding processes are fast, often diffusion-limited,[40,41] with rate constants of the order of 10^7-10^9 M^{-1} s^{-1}. However, if the binding induces a conformational transition the rate constant for the transition can be small, and so the rate of the whole process can be much smaller than that expected for a diffusion-limited process. For this reason we modified the model by introducing a glucose-induced conformational change that would be the limiting step of the reaction, and would be faster when glucose binds to the free enzyme (E_I) than when it binds to the binary complex E_I-MgATP. As the rate of the conformational change would be of similar magnitude to the catalytic constant the pathways of formation of the ternary complex would not be at rapid equilibrium but in steady state, and the relative distribution of the two binary complexes in the steady state would vary with the relative concentrations of the two substrates. The two pathways would be non-equivalent from a kinetic point of view because of the difference in rate in the conformational change induced by glucose. The enzyme liberated at the end of the catalytic cycle (E_{II}) would relax quickly to the conformation E_I, so the free enzyme would exist largely in this conformation at low glucose concentrations.

A key point for understanding this model would be to establish whether the cooperativity arises principally because increasing the glucose concentration increases the fraction of the pathway that operates through the binary complex E_I-glucose, or because increasing the glucose concentration increases the probability that glucose can trap the species E_{II} before it relaxes to E_I, or because of both together.

An extension of this model, which includes all the possible enzyme species, was simulated on an analog computer, using a stochastic method developed by Olavarría.[42,43] The simulated model proved able to account for most of the observations mentioned above.

The steady-state random model, with or without the conformational change, explains well the dependence of the cooperativity on the concentration of MgATP and its suppression by the presence of glucose analogs. (The failure of *N*-acetylglucosamine or mannose to inhibit hexokinase D at low glucose concentration is better explained by the models that include conformational transitions, however).

Nonetheless, difficulties remain. For example, it was afterwards shown that the mechanism is mostly ordered with glucose as first substrate and only a small degree of randomness.[44] I considered, as did Cornish-Bowden's group, that this small degree of randomness was insufficient to generate observable cooperativity and for this reason I finally arrived at a model along the lines of the slow-transition model with ordered binding of substrates.[13,14] However, Pettersson later argued that the cooperativity of hexokinase D could be well explained by a random-order mechanism,[10] pointing out that even if the alternative pathway accounts for a very small proportion of the total flux it can generate easily observable deviations from Michaelis-Menten kinetics. This does not completely solve the problem, because the random-order mechanism predicts that both substrates should show deviations from Michaelis-Menten kinetics, whereas hexokinase D shows no such deviations for MgATP.[45-47] So although it accounts satisfactorily for the behavior as a function of glucose concentration it predicts observable inhibition by MgATP that in fact is not observed. It can, of course, be argued that the terms in $[\text{MgATP}]^2$ were present in the rate equation but were undetectable in the experimental range of concentrations so that the predicted kinetic behavior could not be distinguished from Michaelis-Menten kinetics.

Compulsory-order mechanism with recycling of an enzyme-product complex

Petterson also proposed another model, shown in Figure 6.10, in which there is recycling of an enzyme-product complex via an enzyme-substrate-product complex.[11] Although this proposal was made to account for the kinetics of the wheat-germ hexokinase type L_I it would appear at first sight as an attractive hypothesis to explain the behavior of hexokinase D.[47] The reaction would occur by a compulsory-order mechanism with glucose as a first substrate, but with a random-order release of products. If a binary enzyme-MgADP complex is capable of binding glucose to give a ternary enzyme-glucose-MgADP complex and then releasing MgADP, it can allow enzyme to be recycled that might otherwise be sequestered as a slowly dissociating enzyme-product complex (Fig. 6.10). Such cycling would generate a higher-order dependence

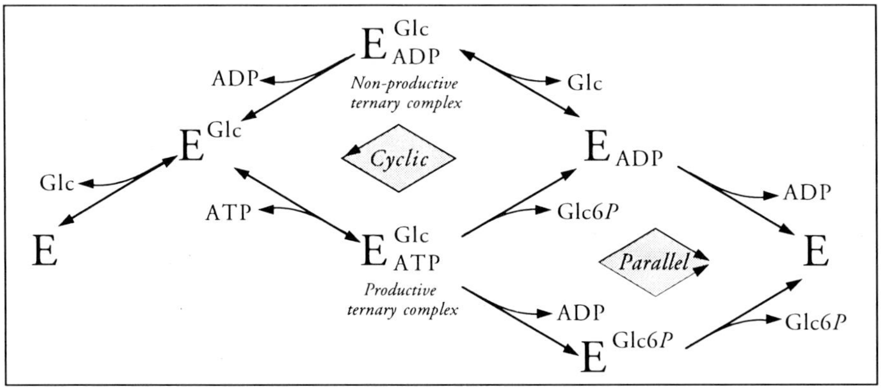

Fig. 6.10. Recycling of an enzyme-product complex. If a substrate can trap an enzyme-product complex to regenerate its binary complex with the enzyme, this can generate cooperativity with respect to the trapping substrate. Note that despite the superficially similar appearance of the two diamond-shaped fragments of the mechanism, the irreversibility of product-release steps ensures that the flux is exclusively cyclic in one and parallel in the other.

on the glucose concentration, and, unlike the random-order mechanism, would imply no higher-order dependence on the MgATP concentration.

One might object that it is improbable that substrate binding follows a compulsory order whereas products can be released in either order; however, this is not an essential feature of the model, and similar properties would occur with random binding of substrates in rapid equilibrium.[47] Nonetheless, a problem is that this model requires an enzyme-MgADP complex, which could not be detected in isotope-exchange measurements.[44] Both trapping experiments and flux-ratio experiments failed to show any evidence for such a complex though there was no difficulty in detecting the existence of the binary enzyme-glucose-6-phosphate complex. On the other hand the fact that MgADP protects hexokinase D against inactivation by DTNB, albeit feebly, argues in favor of its existence.[23]

SLOW-TRANSITION MODEL

Various investigators have discussed models that involve slow isomerization between two forms of an enzyme with a single active site.[40,48-50] In these cases, both forms bind the substrate, but have different affinities or different rate constants for catalysis.

The slow-transition model proposed by Neet's group is the most thoroughly developed of these models and is the most general in character. It includes the following essential postulates: (i) the enzyme can exist in two (or more) different conformational states that are interconvertible at rates that are slow compared with the rate of the catalytic reaction; (ii) each form of the enzyme is capable of accomplishing a complete catalytic cycle, but the rate constants are different; (iii) the rate of interconversion and the proportions of the two states are influenced by the substrate concentration. The binding and isomerization step of the mechanism should be in the steady state.

The steady-state rate for this mechanism is the sum of the rates for the two catalytic cycles. As the substrate concentration is varied the steady-state distribution of the enzyme between the two forms varies also and, provided that they have different kinetic constants, gives rise to cooperativity, which can be either positive or negative. The slow transition required by the model can be an isomerization, an association-dissociation reaction or binding of another ligand (i.e. modifier), as discussed in more detail by Neet and Ainslie.[50]

If the reaction requires two substrates and two products, as in the case of hexokinase D,[13,14] the analysis becomes more complicated, but essentially the same principles apply as to the one-substrate case. Random or ordered binding of the substrates to the enzyme is compatible with this model as long as the binding is not in rapid equilibrium.

Applied to hexokinase D, this model postulates that the enzyme exists mainly in one conformational state (E_I) in the absence of ligands and that the binding of glucose or mannose induces a conformational transition to a new conformation E_{II} (Fig. 6.11). The rate of the transition from E_I to E_{II} is slow in relation to the catalytic step and the two enzyme forms E_I and E_{II} do not reach equilibrium concentrations in the steady state. The new conformation E_{II} has a higher affinity for the sugar substrates and a higher catalytic activity than E_I. Cooperativity arises from shifts of the steady-state distribution between the two enzyme forms as the sugar concentration increases. Inhibitors suppress the cooperativity with respect to glucose by inducing or trapping the E_{II} conformation, i.e. at high enough concentrations of the inhibitors only one conformation (E_{II}) exists, because as E_{II} binds the inhibitors with higher affinity than E_I, they can trap the E_{II} conformation as it is generated. The different capacity of the inhibitors for suppressing cooperativity results from differences in the rate constants of conformational transitions of the different binary complexes E_I. Thus the

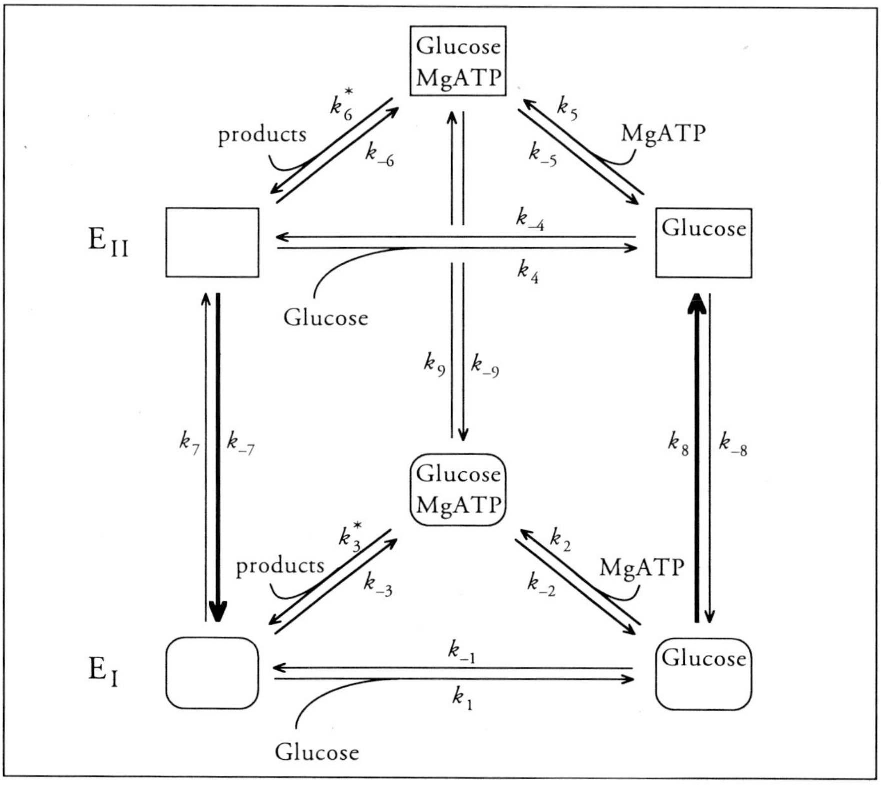

Fig. 6.11. Slow-transition model. In this model there are two conformations E_I (rounded rectangles) and E_{II} (rectangles) of enzyme that exist in all states of substrate binding and are prevented from equilibration by the rapidity of product release from the ternary complexes. As indicated by the thicker arrows, the E_{II} conformation is more stable when glucose is bound and less stable in the free enzyme. As in Figure 6.9 complexities in the order of product release are ignored.

frequency of the conformational transition of the sugar binary complexes would decrease in the following order: Fru > Glc ≈ Man > dGlc > GlcNAc.[13,14,28]

Fructose would be the most efficient for suppressing cooperativity because not only can it trap E_{II}, but it is also very efficient at inducing this conformation, as the transition E_I-Fru→ E_{II}-Fru occurs faster than in the other cases. This high rate constant of conformational transition would also explain the Michaelis-Menten behavior of hexokinase D with respect to fructose. Thus we postulated that with this substrate the reaction would operate mainly through the binary E_{II}-fructose complex, which would explain the higher maximum rate. Consequently only one kinetic pathway would exist and Michaelis-Menten behavior should be expected. In the case of 2-deoxyglucose we also postulated only one kinetic pathway, but in this case through the complex E_I-2-deoxyglucose. The reason for postulating these differences between fructose and 2-deoxyglucose derive from differences in maximal rate with both substrates and the different capacity for suppressing the cooperativity with glucose when used as inhibitors.[13,14,28] *N*-Acetylglucosamine is the least efficient in suppressing cooperativity in spite of its low K_i value because of the low frequency of the conformational transition of the binary complex. Thus, although E_{II} has a high affinity for *N*-acetylglucosamine, explaining the low K_i value (0.11-0.14 mM), this inhibitor can only induce the conformational transition with very low rate, i.e. it is a very poor inducer by itself, but it is very efficient in trapping the E_{II} conformation induced by glucose. If this interpretation is correct one would expect that if *N*-acetylglucosamine is used for protection of the enzyme against inactivation, the dissociation constant obtained would be much bigger than the K_i value, because in this case all the enzyme would be in the E_I conformation and no glucose would be present to induce the formation of E_{II}, which can easily be trapped by *N*-acetylglucosamine. In fact this is what is observed when *N*-acetylglucosamine is used to protect against inactivation by *N*-(*N*-bromoacetyl)-6-aminohexanoylglucosamine, where a dissociation constant of 4.6 mM was obtained,[22] at least 20-fold larger than the kinetic inhibition constant.

The mnemonical model

This model also involves a conformational change during the catalytic cycle and has been the model preferred by Cornish-Bowden's group as the main source of cooperativity in hexokinase D. It originated in a study of Rabin,[51] who described a model for monomeric cooperativity that depends on the existence of two forms of free enzyme differing in stability and isomerizing relatively slowly, with the form of free enzyme released at the end of the reaction being the less stable of the two. (A similar model proposed earlier by Weber and Anderson[52] was open to the objection that it violated the principle of microscopic reversibility, but Rabin's formulation avoided this problem). Whitehead subsequently developed these ideas and pointed out that they imply the existence of "enzyme memory":[53] if the substrate concentration is low, there is ample time for free enzyme to relax to its more stable state, but if it is high substrate can bind to the less stable form before the enzyme "forgets" its earlier interaction with substrate. Ricard introduced the name *mnemonical model* for this type of mechanism.[54] He analyzed it very thoroughly and extended it to apply to reactions of two substrates, as he was working at the time with wheat hexokinase, a monomeric enzyme that shows negative cooperativity with respect to glucose.[55]

The essential postulates of the mechanism are as follows: (i) the enzyme can exist in two interconvertible conformational states, with an equilibrium constant that greatly favors one state in the absence of ligands; (ii) the free enzyme liberated at the end of the catalytic cycle is the less stable form (and is thus different from the form predominant initially); (iii) binding of the substrate (or of the first substrate when there are more than one) to either form of the free enzyme induces a new conformation necessary for catalysis; (iv) the rate of the conformational change induced by the substrate is different for the two states of free enzyme. For a two-substrate reaction the model implies that the substrates bind in a definite order, and cooperativity is expected with respect to the first substrate only. A small degree of randomness in the order of binding, as we believe to occur with hexokinase D, does not eliminate cooperativity entirely, but it tends to decrease it (as well as generating much more complicated rate equations). In this sense, therefore, there is an important difference between the mnemonical mechanism and the slow-transition model, in which randomness of substrate binding can be incorporated without any major

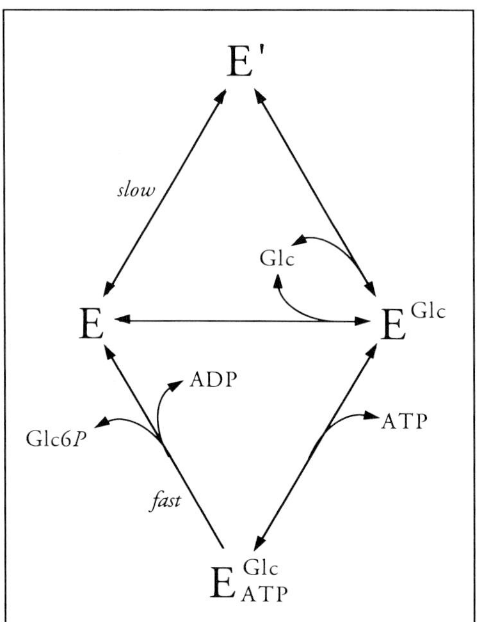

Fig. 6.12. The mnemonical model. Two forms of free enzyme bind glucose to form the same binary enzyme-glucose complex, which is capable of binding MgATP and breaking down to products sufficiently fast to prevent equilibration of the two forms.

effect on the degree of cooperativity. Another difference is that only one catalytic cycle occurs in the mnemonical mechanism whereas the slow transition mechanism has two catalytic cycles.

Storer and Cornish-Bowden applied this model, shown in Fig. 6.12, to the kinetic behavior of hexokinase D with respect to glucose.[56] The two different states of free enzyme E and E' can both react with glucose to form the same enzyme-glucose complex, but the catalytic cycle leads exclusively to the less stable form, E, which relaxes slowly to E'. As both E and E' produce the same enzyme-glucose complex, it follows from thermodynamic considerations that the less stable form E must bind glucose more tightly than E'. In the proposed mechanism MgATP binds after glucose, and this is a necessary feature if one is to account for the effect of MgATP on the glucose cooperativity. At low MgATP concentrations the binding of MgATP to the binary enzyme-glucose complex becomes slower and when its concentration is sufficiently low this step becomes slow enough for the two forms of free enzyme to equilibrate with one another and with glucose; under these circumstances, therefore, one must expect the normal behavior for binding at a single site at equilibrium, i.e. no cooperativity. As the MgATP concentration rises, however, it becomes possible for the enzyme-glucose complex to be removed so fast that it cannot achieve equilibrium, and under these conditions one is no longer constrained by the rules governing equilibria. If the glucose concentration is low, E can relax to E' and one will observe the weak glucose binding characteristic of E', but if it is high it can react with E before E can relax to E', and one will observe the stronger binding characteristic of E; thus at high MgATP concentrations the binding of glucose will appear to become tighter as the glucose concentration increases, and there will be kinetic cooperativity.

According to this model variation of the concentration of MgATP may alter the magnitude of the cooperativity but it cannot alter its sign; it is impossible to pass from positive cooperativity at one MgATP concentration to negative cooperativity at another if the other conditions are unchanged. This may not necessarily be true for other models of kinetic cooperativity, as suggested by simulation studies,[57] and may therefore provide an experimental method for discriminating between mechanisms.

The effect of MgATP on the glucose cooperativity is only possible in the way described if the catalytic reaction that follows MgATP binding is capable of proceeding very fast; if this step were rate-limiting it would not be possible to prevent equilibration of glucose binding by raising the MgATP concentration. Measurements of isotopic exchange at equilibrium showed that exchange between ATP and ADP was much faster than between glucose and glucose 6-phosphate under all conditions;[44] it is not possible, therefore, for both exchanges to proceed through the same rate-limiting step and so the catalytic step of the reaction must be fast, as proposed.

In the presence of MgITP, an analog of MgATP that acts as a poor substrate (Chapter 3), the positive cooperativity with respect to glucose disappears. This may be rationalized in relation to kinetic models for hexokinase D cooperativity, which assume that with the normal substrates the chemical reaction and subsequent release of products occur so rapidly that binding of substrates cannot approach equilibrium and is therefore not constrained by the thermodynamic requirement that the Hill coefficient for substrate binding cannot exceed the number of binding sites. MgITP is a much poorer substrate than MgATP, however; its apparent K_m value at high glucose concentrations is 24 times that of MgATP, whereas the value of the limiting rate V is decreased about 8-fold. This means not only that MgITP binds more weakly than MgATP, which could be overcome by raising the concentration, but also that with MgITP there is a slower chemical reaction or slower release of products, so that it is no longer possible for the ternary complex to be converted into products rapidly enough to generate kinetic cooperativity. The negative cooperativity with respect to glucose observed in ^2H_2O with MgATP as phosphate donor also disappears when MgITP is used instead of MgATP.

The most important feature of the mnemonical model and the slow-transition model is the existence of two different forms of free enzyme with different properties. This feature was difficult to test at the time that these models were proposed, because there was insufficient enzyme to investigate protein structure in any detail. Now that the enzyme has been cloned and can be produced in large quantities protein structure studies are possible. For example, mutagenesis could allow production of an enzyme with decreased activity, with the catalytic step slow enough to allow equilibration between the two pathways with elimination of the cooperativity.

How slow is slow?

As both the slow-transition and mnemonical models depend on the presence of slow steps in the mechanism, it is useful to pause briefly to consider what we mean by "slow." In fact, the clumsier terms "not fast" or "intermediate rate" would be more accurate, because in both cases we do not mean processes that are very slow compared with the time scale of the experiment, such as relaxations with half times exceeding 30 minutes. Such processes would be too slow to generate cooperativity that appeared to be instantaneous in the normal assays. In the context of these models, therefore, a slow process is one with a time constant similar in magnitude to that of the catalytic process - not much faster, as it would be completed before the catalytic reaction occurred to any significant extent, and not much slower, as then it would either not be observed at all during the assay or would be observed as a gradual change in enzyme activity during the experiment.

EVIDENCE FOR DIFFERENT CONFORMATIONS

Some direct evidence for the existence of different conformational states exists, but it is not extensive. Measurements of the native intrinsic fluorescence of hexokinase D have provided useful information, and a glucose-induced conformational change in the hexokinase D molecule has been observed (Fig. 6.4).[14,24] Enhancement of intrinsic fluorescence occurs slowly on glucose binding (Fig. 6.13), with a rate that increases with the glucose concentration;[24] this is consistent with slow isomerization between two forms of hexokinase D with different glucose affinities. As in the kinetic experiments the affinity for glucose increases in the presence of glycerol,[14,24] and the rate of the transition decreases.[24] The half times for the kinetic assay transient and the fluorescent conformational change are both between 0.5 and 5 min, and they have similar dependences on glucose concentration, thus providing strong evidence that the basis for the cooperativity of hexokinase D is a ligand-induced slow transition.[24]

We had postulated that 2-deoxyglucose would be unable to induce the high-affinity form induced by glucose,[13] as mentioned above. If this were the case the titration of hexokinase D with 2-deoxyglucose should produce no fluorescence enhancement or would produce it much more slowly than glucose. Neet's group did this experiment and

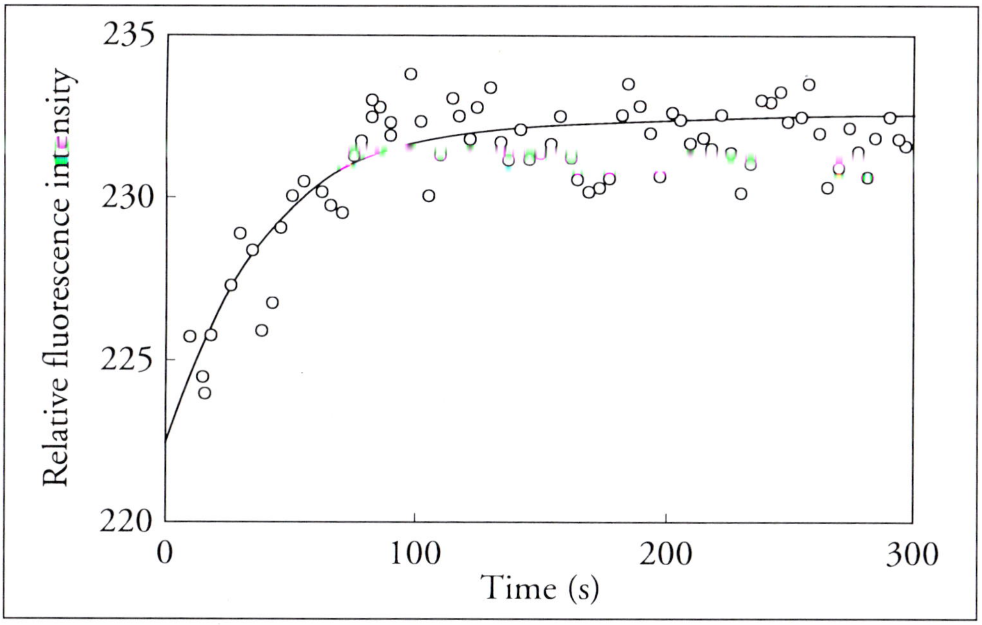

Fig. 6.13. Direct evidence for a slow conformational change in hexokinase D. The figure shows the slow change in fluorescence that occurs when glucose binds to the enzyme. Reproduced with permission from J Biol Chem 1990; 265:9670-9675.

observed no enhancement, but a rapid fluorescence quenching of at least 8%.[24] One may speculate that the glucose-induced conformational change of hexokinase D seen in the change of fluorescence is the same as the domain movement and cleft closure seen in crystallographic studies of yeast hexokinase.[58,59] Yeast hexokinase also shows a fluorescence change,[60] but the isomerization appears to be rapid for this non-cooperative enzyme.[24] In addition, the intrinsic fluorescence change of the yeast enzyme on binding of glucose is an 8% quenching as opposed to the 15-20% enhancement seen with hexokinase D.[14,24] The sequences of these two enzymes are now known to be about 33% identical, with another 30% of similar residues,[20] and hexokinase D has been cloned and is readily expressed in bacteria. Determination of the amino acid substitutions that cause these kinetic and functional differences will be of great interest.

Glycerol added to assays depresses cooperativity,[14,32] as mentioned above, and decreases the rate of the molecular transition sufficiently for a slow transient to be observed in reaction assays.[61] Consistent with effects on an enzyme isomerization, there was a burst under these conditions if the enzyme was preincubated at high glucose concentration

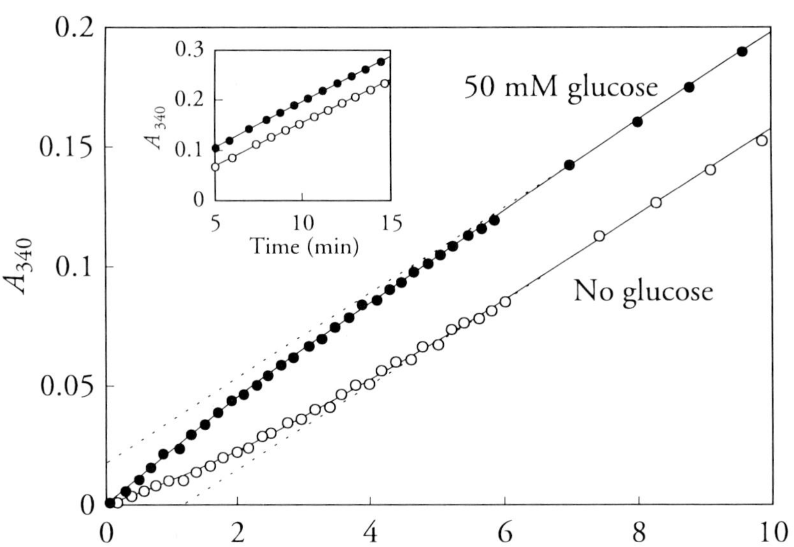

Fig. 6.14. Progress curves in the presence of glycerol. Assay conditions were identical for the two curves, but there was a lag if all glucose was removed by dialysis from the enzyme stock solution, whereas there was a burst if the enzyme was preincubated with 50 mM glucose. The curves are extended to a total time of 15 min in the inset, from which it is evident that the steady-state rates were essentially the same in the two cases. Redrawn from Biochemistry 1990; 29:770-777.

in the presence of glycerol, but there was a lag if it was initially kept at low glucose concentration or dialyzed before the assay (Fig. 6.14). The steady-state rates after the transient showed the cooperativity and other features of the linear hexokinase D assay without glycerol.[61]

CURRENT STATUS OF MODELS FOR HEXOKINASE D COOPERATIVITY

Although the large degree of overlap between the predictions made by different models and the lack of experimental data in some crucial areas means that it is not yet possible to make a definitive choice of mechanism to explain the cooperativity of hexokinase D, some important conclusions can still be drawn. First of all there can be no doubt that the cooperativity is kinetic in origin and that it cannot be explained by a variant of one of the classic mechanisms. In this respect hexokinase D remains an outsider among enzymes showing a significant degree of positive cooperativity; virtually all other such enzymes are oligomers and their cooperativity is easily explained in terms of subunit interactions. The second point is that there is good evidence, as we have seen in this chapter, for a mechanism involving slow processes in the catalytic cycle, but whether this is of the slow-transition or mnemonical type is less clear. In general the bulk of the evidence favors a slow-transition mechanism[6] of the type proposed originally by Neet's group, and I shall conclude by summarizing this evidence.

The slow-transition model accounts for various properties described in Chapter 3: sigmoidal kinetics with respect to glucose and mannose but Michaelis-Menten kinetics with respect to 2-deoxyglucose and fructose; suppression of sigmoidicity and increase of affinity for glucose at low MgATP concentrations and hyperbolic behavior with respect to this substrate. It also explains the suppression of cooperativity with respect to glucose in the presence of competitive inhibitors and the cooperative effects of some of these inhibitors at low glucose concentrations. It is consistent with the effects of the inhibitors *N*-acetylglucosamine[12,13,61] and palmitoyl-CoA[62] on cooperativity,[63] and with the negative cooperativity observed with MgATP[61] or glucose[12,57,61] under certain conditions. It also accounts for some regulatory properties that will be dealt with in Chapter 7: partial inhibition by palmitoyl-CoA without modification of the cooperativity with respect to glucose; complete inhibition by the binary complex between the regulatory protein and fructose 6-phosphate, with slight diminution of cooperativity.

PERSPECTIVES

In 1980 Neet and Ainslie proposed four structural types of conformational changes that could be slow enough to account for hysteretic mechanisms,[50] namely (i) rotation or sliding of α-helices with respect to each other; (ii) *cis-trans* isomerization of proline residues, which is known to occur, though slowly; (iii) "segmental rotation" of a relatively large portion of the protein; and (iv) a "hinged movement" of domains (cleft closure).

Each of these changes has the potential for a high activation energy.[6] Cleft closure is made highly probable for hexokinase D by the observation of the glucose-induced domain movement of the homologous yeast hexokinase,[59] but there are no direct data for hexokinase D itself. Proline isomerization is certainly involved in slow steps in protein folding,[64] but it appears that no ligand-induced isomerization of proline has been documented. This would certainly be something to explore in the case of hexokinase D. Standard x-ray crystallographic methods provide additional possibilities for describing the conformational change induced by glucose, but little can be said at the moment about the kinetics of such slow changes.

According to Neet,[6] a method with good prospects is the Laue diffraction method, now being used to study time-dependent changes during catalysis.[65,66] It could be used in future studies to describe slow conformational changes at high resolution.

Exploration of all of these approaches and others has been made possible by the availability of hexokinase D in large amounts, as both the native enzyme and mutants with selected properties.

References

1. Hue L, Hers H-G. Utile and futile cycles in the liver. Biochem Biophys Res Comm 1974; 58:540-548.
2. Monod J, Changeux J-P, Jacob F. Allosteric proteins and cellular control systems. J Mol Biol 1963; 6:306-329.
3. Monod J, Wyman J, Changeux J-P. On the natural of allosteric transitions. A plausible model. J Mol Biol 1965; 12:88-112.
4. Koshland DE Jr, Némethy G, Filmer D. Comparison of experimental binding data and theoretical models in proteins containing subunits. Biochemistry 1966; 5:365-385.
5. Ricard J, Cornish-Bowden A. Co-operative and allosteric enzymes: 20 years on. Eur J Biochem 1987; 166:255-272.
6. Neet KE. Cooperativity in enzyme function: equilibrium and kinetic aspects. Meth Enzymol 1995; 249:519-567.
7. Ferdinand W. The interpretation of non-hyperbolic rate curves for two-substrate enzymes. A possible mechanism for phosphofructokinase. Biochem J 1966; 98:278-283.
8. Cornish-Bowden A, Cárdenas ML. Co-operativity in monomeric enzymes. J Theor Biol 1987; 124:1-23.
9. Ricard J. Generalized microscopic reversibility, kinetic co-operativity of enzymes and evolution. Biochem J 1978; 175:779-791.
10. Pettersson G. Mechanistic origin of the sigmoidal behaviour of glucokinase. Biochem J 1986; 233:347-350.
11. Pettersson G. Mechanistic origin of the kinetic cooperativity of hexokinase type L_1 from wheat germ. Eur J Biochem 1986; 154:167-170.

12. Cárdenas ML, Rabajille E, Niemeyer H Kinetic cooperativity of glucokinase with glucose. Arch Biol Med Exp 1979; 12:571-580.
13. Cárdenas ML, Rabajille E, Niemeyer, H. Suppression of kinetic cooperativity of hexokinase D ("glucokinase") by competitive inhibitors. Eur J Biochem 1984; 145:163-171.
14. Cárdenas ML, Rabajille E Trayer IP et al. Cooperative interactions in hexokinase D ("glucokinase"). Kinetic and fluorescence studies. Arch Biol Med Exp 1985; 18:273-284.
15. Davies GE, Stark GR. Use of dimethyl suberimididate, a cross-linking reagent, in studying the subunit structure of oligomeric proteins. Proc Natl Acad Sci USA 1970; 66:651-656.
16. Holroyde MJ, Allen MB, Storer AC et al. The purification in high yield and characterization of rat hepatic glucokinase. Biochem J 1975; 153:363-373.
17. Cárdenas ML, Rabajille E, Niemeyer H. Maintenance of the monomeric structure of glucokinase under reacting conditions. Arch Biochem Biophys 1978; 190:142-148.
18. Grossman SH, Dorn CG, Potter VR. The preparation and purification of pure rat liver glucokinase. J Biol Chem 1974; 249:3055-3060.
19. Maccioni R. Purificación y caracterización de glucoquinasa. Doctoral thesis, University of Chile, 1975.
20. Andreone TL, Printz R.L, Pilkis SJ et al. The amino acid sequence of rat liver glucokinase deduced from cloned cDNA. J Biol Chem 1989; 264:363-369.
21. Hayzer DJ, Iynedjian PB. Alternative splicing of glucokinase mRNA in rat liver. Biochem J 1990; 270:261-263.
22. Connolly BA, Trayer IP. Reaction of rat hepatic glucokinase with substrate-related and other alkylating agents. Eur J Biochem 1979; 99:299-308.
23. Monasterio O, Orellana O, Heberlein U et al. Protective effect of various ligands against glucokinase inactivation by 5,5'-dithiobis-2-nitrobenzoic acid. Arch Biol Med Exper 1981; 14:139-142.
24. Lin SX, Neet KE. Demonstration of a slow conformational change in liver glucokinase by fluorescence spectroscopy. J Biol Chem 1990; 265:9670-9675.
25. González C, Ureta T, Babul J et al. Characterization of isoenzymes of adenosine triphosphate:D-hexose 6-phosphotransferase from rat liver. Biochemistry 1967; 6:460-468.
26. Salas J, Viñuela E, Salas M et al. Glucokinase of rabbit liver. Purification and properties. J Biol Chem 1965; 240:1014-1018.
27. Parry MJ, Walker DG. Purification and properties of adenosine 5'-triphosphate-D-glucose 6-phosphotransferase from rat liver. Biochem J 1966; 99:266-274.
28. Cárdenas ML, Rabajille E, Niemeyer H. Fructose: a good substrate for rat-liver "glucokinase" (hexokinase D). Biochem J 1984; 222:363-370.
29. Jarabak J. Human placental 15-hydroxyprostaglandin dehydrogenase. Proc Natl Acad Sci USA 1972; 69:533-534.
30. Ruwart MJ, Suelter CH. Activation of yeast pyruvate kinase by natural and artificial cryoprotectants. J Biol Chem 1971; 246:5990-5993.

31. Tippett PS, Neet KE. Interconversions between different sulfhydryl-related kinetic states in glucokinase. Arch Biochem Biophys 1983; 222:285-298.
32. Pollard-Knight D, Connolly BA, Cornish-Bowden A et al. Effect of glycerol on glucokinase activity: loss of co-operative behavior with respect to glucose. Arch Biochem Biophys 1985; 237:328-334.
33. Gekko K, Timasheff SN. Mechanism of protein stabilization by glycerol: preferential hydration in glycerol-water mixtures. Biochemistry 1981; 20:4667-4676.
34. Gekko K, Timasheff SN. Thermodynamic and kinetic examination of protein stabilization by glycerol. Biochemistry 1981; 20:4677-4686.
35. Pollard-Knight DV, Cornish-Bowden A. Solvent isotope effects on the glucokinase reaction. Negative co-operativity and a large inverse isotope effect in ^2H_2O. Eur J Biochem 1984; 141:157-163.
36. Wong JT, Hanes CS. Kinetic formulations for enzymic reactions involving two substrates. Canad J Biochem Physiol 1962; 40:763-804.
37. Dixon M, Webb EC. In: Enzymes. 3rd ed. New York: Academic Press. 1979:88-90.
38. Niemeyer H, Monasterio O, Cárdenas ML et al. Kinetic mechanism of the phosphorylation of 2-deoxyglucose by rat liver glucokinase. Fed Proc 1978; 37:1421.
39. Colowick SP. The hexokinases. In: Boyer PD, ed. The Enzymes. 3rd ed. New York: Academic Press, 1973; 9:1-48.
40. Frieden C. Kinetic aspects of regulation of metabolic processes. The hysteretic enzyme concept. J Biol Chem 1970; 245:5788-5799.
41. Gutfreund H. Enzymes: physical principles. Wiley, 1975.
42. Olavarría JM. Simulación directa de sistemas enzimáticos, utilizando un sistema estocástico de reacción enzima-sustrato. An Acad Cienc Exact Fís Nat 1973; 25:129-165.
43. Olavarría JM. Simulación directa. Rev. Inst. Cibernética Soc. Cient Argent 1980; 5:88-102.
44. Gregoriou M, Trayer IP, Cornish-Bowden A. Isotope exchange evidence for an ordered mechanism for rat-liver glucokinase, a monomeric co-operative enzyme. Biochemistry 1981; 20:499-506.
45. Storer AC, Cornish-Bowden A. Kinetics of rat liver glucokinase. Co-operative interactions with glucose at physiologically significant concentrations. Biochem J 1976; 159:7-14.
46. Cárdenas ML. Glucoquinasa, una enzima monomérica con cinética cooperativa. Doctoral thesis, University of Chile, 1982.
47. Cornish-Bowden A, Storer AC. Mechanistic origin of the sigmoidal rate behaviour of rat liver hexokinase D ("glucokinase"). Biochem J 1986; 240:293-296.
48. Witzel H. Zur Interpretation kinetischer Daten der Reaktion der Pancreas-Ribonuclease. Hoppe-Seylers Z Physiol Chem 1967; 348:1249-1250.
49. Ainslie GR, Shill JP, Neet KE. Transients and cooperativity. A slow transition model for relating transients and cooperative kinetics of enzymes. J

Biol Chem 1972; 247:7088-7096.

50. Neet KE, Ainslie GR. Hysteretic enzymes. Meth Enzymol 1980; 64:192-226.
51. Rabin BR. Cooperative effects in enzyme catalysis: a possible kinetic model based on substrate-induced conformation isomerization. Biochem J 1967; 102:22c-23c.
52. Weber G, Anderson SR. Multiplicity of binding. Range of validity and practical test of Adair's equation. Biochemistry 1965; 4:1942-1947.
53. Whitehead E. The regulation of enzyme activity and allosteric transition. Prog Biophys Mol Biol 1970; 21:321-397.
54. Ricard J, Meunier J-C, Buc J. Regulatory behavior of monomeric enzymes. 1. The mnemonical enzyme concept. Eur J Biochem 1974; 49:195-208.
55. Meunier J-C, Buc J, Navarro A. et al. Regulatory behavior of monomeric enzymes. 2. A wheat-germ hexokinase as a mnemonical enzyme. Eur J Biochem 1974; 49:209-223.
56. Storer AC, Cornish-Bowden A. Kinetic evidence for a "mnemonical" mechanism for rat liver glucokinase. Biochem J 1977; 165:61-69.
57. Olavarría JM, Cárdenas ML. Cooperative behaviour in monomeric enzymes. Change of negative to positive cooperativity by effect of a ligand. Arch Biol Med Exper 1985; 18:285-292.
58. Steitz TA, Anderson WF, Fletterick RJ et al. High-resolution crystallographic structures of yeast hexokinase complexes with substrates, activators, and inhibitors. Evidence for an allosteric control site. J Biol Chem 1977; 252:4494-4500.
59. Bennett WS, Steitz TA. Structure of a complex between yeast hexokinase A and glucose. II. Detailed comparisons of conformation and active site configuration with the native hexokinase B monomer and dimer. J Mol Biol 1980; 140:211-230.
60. Ohning GV, Neet KE. 6-(*p*-Toluidinyl)naphthalene-2-sulfonic acid as a fluorescent probe of yeast hexokinase. Conformational states induced by sugar and nucleotide ligands. Biochemistry 1983; 22:2986-2995.
61. Neet KE, Keenan RP, Tippett PS. Observation of a kinetic slow transition in monomeric glucokinase. Biochemistry 1990; 29:770-777.
62. Tippett PS, Neet KE. An allosteric model for the inhibition of glucokinase by long chain acyl coenzyme A. J Biol Chem 1982; 257:12846-12852.
63. Neet KE, Tippett PS, Keenan RP. Regulatory properties of glucokinase. In: Chock PB, Huang CY, Tsou CL et al., eds. Enzyme Dynamics and Regulation eds. New York: Springer-Verlag, 1988:28-38.
64. Wood LC, White TB, Ramdas L et al. Replacement of a conserved proline eliminates the absorbance-detected slow folding phase of iso-cytochrome c. Biochemistry 1988; 27:8562-8568.
65. Stoddard BL, Koenigs P, Porter N et al. Observation of the light-triggered binding of pyrone to chymotrypsin by Laue x-ray crystallography. Proc Natl Acad Sci USA 1991; 88:5503-5507.
66. Johnson, LN. Time-resolved protein crystallography. Protein Sci 1992; 1:1237-1243.

CHAPTER 7

SHORT-TERM REGULATION: THE REGULATORY PROTEIN

This chapter describes efforts to find a short-term regulatory mechanism for hexokinase D. It describes the allosteric inhibition by long chain acyl-CoA, and gives a detailed account of the discovery of the regulatory protein—one of the most interesting discoveries in the hexokinase D field in recent years—explaining its mechanism of action and some of its characteristics.

SEARCHING FOR A SHORT-TERM REGULATORY MECHANISM

Since the discovery of the control of hexokinase D activity by substrate concentration, and for a long time afterwards, this represented the only type of short-term regulation considered to play a physiological role. However in the minds of several researchers in the field, including myself, there was the idea that the activity of hexokinase D ought to be rapidly regulated by additional means such as covalent modification or allosteric effects, and quite possibly by both. It seemed to be too important a reaction to be regulated in the short term by just one mechanism without any more complex pattern of regulation, contrasting with the common observation that enzyme activities are regulated through several mechanisms operating in a coordinate manner.[1] However my efforts and those of others to find an effector that acted on the purified enzyme failed for a long while, apart from the action of long-chain fatty acyl-CoAs,[2] which was then of arguable physiological importance, as discussed below.

Despite failing to find an effector for the purified enzyme, we had some observations on the time course of activation of the enzyme in crude extracts—the odd, irreproducible type of result that every experimenter has but which would not convince anyone else—suggesting the possibility of covalent modification or the existence of a regulatory factor. The idea of regulation by phosphorylation and dephosphorylation was very appealing, and we thought that if this occurred phosphorylation would probably inactivate hexokinase D, as it does glycogen synthetase, and that the modified enzyme could become sensitive to effectors that had no effect on the purified (unmodified) enzyme. On the other hand, Niemeyer's group had shown by immunochemical titration in rat liver extracts that the decrease in hexokinase D activity in the livers of fasted or diabetic animals was due to a decrease in the amount of enzyme and not to a decrease in the specific activity of hexokinase D.[3] This did not favor the existence of a covalent modification. However, it could be argued that covalent modification or an effector in vivo that could not be detected in extracts might still contribute to inactivate the enzyme, as well as a mechanism that would operate in terms of gene expression. As a matter of fact, as discussed below, glucose phosphorylation in hepatocytes does show different kinetic characteristics from those seen in diluted extracts or with the purified enzyme, and consequently the possibility of an effector or a covalent modification was always latent.

With this in mind Niemeyer made some studies looking for conditions that could reveal previously undetected mechanisms of modulation of hexokinase D. The idea was the opposite to Racker's classic aphorism, "First purify, then think; do not waste clean thoughts on dirty enzymes," which Weissmann[4] reported that he inscribed on the wall of his laboratory, but did not follow. As efforts to find an allosteric regulator of the purified enzyme had failed, Niemeyer studied a "dirty enzyme," concentrated liver cytosol prepared very swiftly from rats subjected to various dietary and hormonal manipulations. As the conditions were closer to those of the cell than those produced by the usual purification procedures they would permit detection of an enzyme susceptible to allosteric effectors or covalent modification. He appeared to have found an inhibitory protein that would act through fructose 2,6-bisphosphate;[5] that factor turned out to be phosphofructokinase, however, and the regulatory effect was an artifact derived from the type of assay.[6] Nevertheless, a regulatory protein does exist, but it was for Van Schaftingen,[7] working with isolated hepatocytes, to discover it a little later, as described below. It exerts its action on hexokinase D not with fructose 2-6-bisphosphate, but through the interplay of fructose 1-phosphate and fructose 6-phosphate.

Phosphorylation of hexokinase D has still not been proved to occur in vivo. It has been reported, however, that purified hexokinase D from rat liver (specific activity 40 units/mg, about 25% of the highest specific activities that have been observed) appears to serve as a substrate for cAMP-dependent protein kinase in vitro,[8] producing a small decrease in its activity. Prolonged incubation with the protein kinase resulted in exclusive phosphorylation of a serine residue, amounting to 1 mol of phosphate per mol of enzyme; the authors estimated that the phosphorylated hexokinase D showed about a 40% decrease in V_{max} and about a 25% increase in the $K_{0.5}$ for glucose. Nevertheless, these numbers have to be taken with caution, as the paper was unclear about how they were obtained, the complex kinetics of hexokinase D with respect to glucose being certainly ignored. In any case the effect on the activity is too modest to have much physiological significance. The smallness of the effect cannot be ascribed to the assay conditions, such as the pH of 8.0 or the high K^+ concentration, as the purified enzyme has essentially the same kinetic characteristics over the pH range 7.2-8.0 and for K^+ concentrations from 0 mM (ionic strength 111 mM) to 150 mM (ionic strength 261 mM), with only a slight variation of V_{max} and $K_{0.5}$ in the latter case, as seen in Chapter 3.

This report[8] contrasts with unsuccessful attempts by Ian Trayer and his colleagues to phosphorylate pure hexokinase D in vitro with cAMP-dependent protein kinase, or to detect the presence of any phosphate in the purified enzyme (I. P. Trayer, personal communication). As it is now well known that it is difficult to completely eliminate the phosphate from many proteins phosphorylated in vivo, this also argues against phosphorylation of hexokinase D in vivo.

Furthermore, although the minimal motif Arg-Xaa-Ser, which can serve as the substrate site for in vitro phosphorylation by cAMP-dependent protein kinase, occurs twice in the aminoacid sequence of rat liver hexokinase D, at positions 358-360 and 394-396, neither of the target motifs Arg-Arg-Xaa-Ser or Arg-Arg-Ser found in proteins phosphorylated in vivo by this kinase[9] are present in hexokinase D.[10] This raises further doubts about the existence of this mechanism in vivo. In addition, Iynedjian has tried to see if a phosphorylated enzyme can be found in certain hormonal or nutritional conditions, by looking for a charge shift in hepatic hexokinase D, again without success,[10] and again in agreement with the results of Niemeyer's[3] and Trayer's groups mentioned above.

All these data taken together make it very unlikely that rat liver hexokinase D is phosphorylated in vivo.

REGULATION BY LONG-CHAIN ACYL DERIVATIVES OF COENZYME A

PALMITOYL-COA INHIBITION OF HEXOKINASE D: A CONTROVERSIAL OBSERVATION

Palmitoyl-CoA was long ago reported as an inhibitor of hexokinase D,[2] and this appeared to have interesting implications in vivo, as hepatic palmitoyl-CoA concentrations had been reported to rise after starvation and production of alloxan diabetes.[11] Both of these are conditions in which gene expression of hexokinase D is decreased, and so inhibition by palmitoyl-CoA would contribute to a further decrease of hexokinase D activity in the same conditions. The physiological role of any inhibition by palmitoyl-CoA was rather controversial, however,[12,13] particularly in view of the surface-active properties of these compounds which are particularly important at concentrations greater than 0.1 mM.

At that time, therefore, it was difficult to make a distinction between a specific, relevant interaction of a long chain acyl-CoA with an enzyme and nonspecific detergent effects of the amphiphile on an enzyme; probably many of the reported cases were nonspecific. Furthermore, it had been reported that long-chain acyl-CoAs inhibit numerous unrelated enzymes in diverse metabolic pathways, without any possible physiological explanation, and that these powerful detergents often inactivated the enzymes irreversibly or inhibited them in a manner that was not characteristic of classical site-specific inhibition. On the other hand, long chain acyl-CoAs (especially palmitoyl- and oleoyl-CoAs) are in a very appropriate metabolic position to act on enzymes of different metabolic pathways, and it should not be surprising to have several different enzymes inhibited by these types of compound.

The fact that the inhibition of hexokinase D by palmitoyl-CoA was reversible in both supernatant and partially purified preparations,[2] instantaneous and observable at concentrations as low as 10 μM, suggested that this inhibition was unlikely to be a nonspecific detergent mechanism, but the doubts persisted.

Specificity and Kinetic Characteristics of the Inhibition by Palmitoyl-CoA

Neet's group reexamined this problem a decade afterwards, and analyzed the effect of different acyl-CoAs.[14] In general, the longer the hydrocarbon chain, the smaller the inhibition constant, and so the stronger the inhibition (Table 7.1). Hexokinase D proved to be specifically

Table 7.1. Apparent inhibition constants and critical micelle concentrations for acyl-coenzymes A[1]

Effector	Apparent inhibition constant[2]	Critical micelle concentration[3] in 5 mM $MgCl_2$	Critical micelle concentration[3] in 10 mM $MgCl_2$
Lauroyl-CoA (C-12)	14.1 μM	67 μM	28.8 μM
Myristoyl-CoA (C-14)	5.5 μM	17 μM	5.8 μM
Palmitoyl-CoA (C-16)	1.8 μM	4.55 μM	1.52 μM
Stearoyl-CoA (C-18)	0.32 μM	1.0 μM	0.4 μM
Oleoyl-CoA (C-18:1)	0.77 μM	6.8 μM	1.2 μM
Palmitoyl carnitine	30 μM	19 μM	19 μM
Palmitic acid	> 1 mM	2 mM	
Myristic acid	> 1 mM	6 mM	
Acetyl-CoA	> 10 mM	—	—
Coenzyme A (free SH)	> 1 mM	—	—

1 Data of Tippett and Neet.[14]

2 Effector concentration needed to double the half-saturation concentration for glucose in the presence of 5 mM ATP and 5 mM $MgCl_2$.

3 Determined in mixtures with pinacyanol chloride and the concentrations of $MgCl_2$ shown.

inhibited by the long chain fatty acyl-CoAs, but not by related free fatty acid anions such as myristate or palmitate at concentrations as high as 1 mM, nor by short-chain acyl-CoAs, free CoASH or similar amphiphiles. Palmitoyl carnitine does inhibit hexokinase D (Table 7.1), although it is not a long chain fatty acyl-CoA; however, it inhibited only at concentrations above 20 μM, and, more importantly, the inhibition is time-dependent and irreversible. Furthermore, the critical micelle

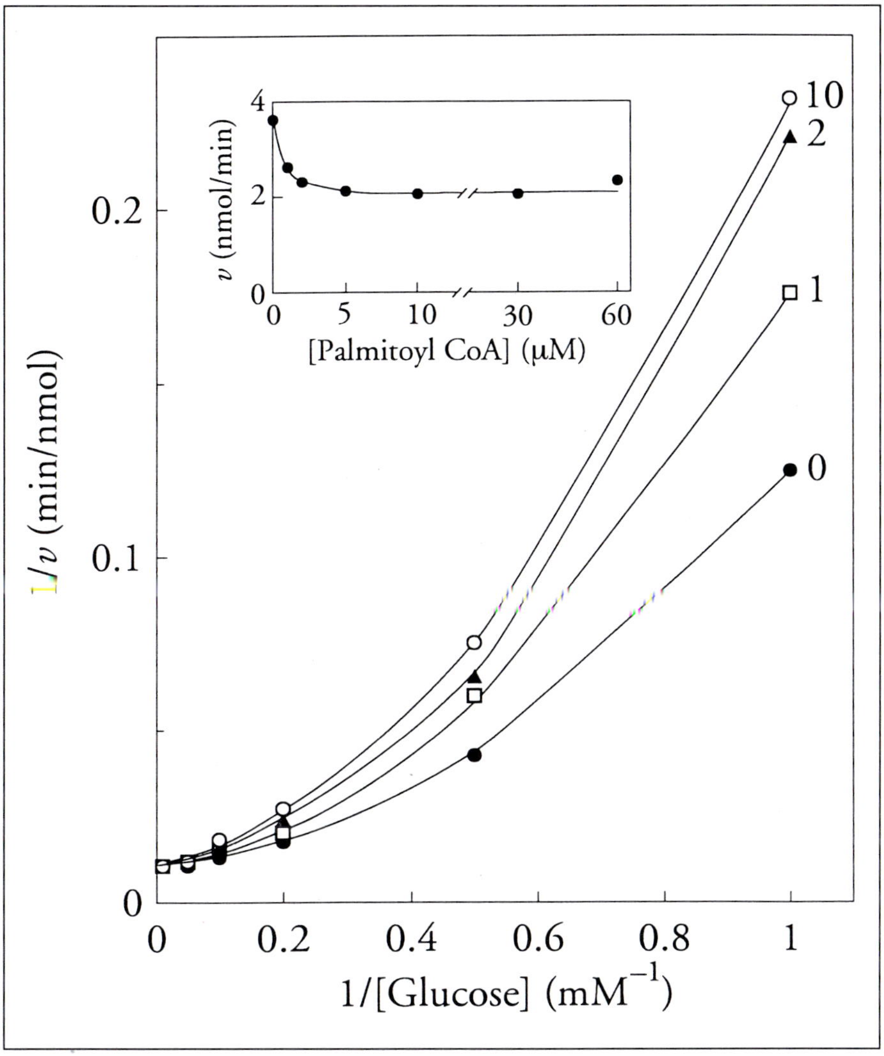

Fig. 7.1. Competitive inhibition of hexokinase D by palmitoyl-CoA at different glucose concentrations, 5 mM total ATP and 10 mM total Mg^{2+}. The effect of Mg^{2+} on the results is discussed below in the text. The inset shows the activity as a function of the palmitoyl-CoA concentration at 2 mM glucose. Note that the inhibition is partial, i.e. that the rate does not approach zero at saturation. Redrawn from J Biol Chem 1982; 257:12839-12845.

concentration is 19 μM in assays with either low or high magnesium concentrations, and one can conclude, as did the authors, that this inhibition is probably due to nonspecific detergent action of palmitoyl carnitine on hexokinase D. In contrast to palmitoyl carnitine, the inhibitory effect of long chain acyl-CoAs on hexokinase D is immediate, instantly reversible and occurs with concentrations of fatty acyl-CoAs lower than the critical micelle concentration.[14] On this basis the authors reasonably argued that the inhibition reflects specific binding of the fatty acid derivatives to the enzyme, and not a trivial detergent effect or an unspecific lipid-protein interaction.

Furthermore, kinetic studies have shown that palmitoyl-CoA acts as a competitive inhibitor of hexokinase D with respect to glucose (Fig. 7.1) and, curiously, also with MgATP (it decreases the binding affinity of glucose and MgATP without impairing V_{max}). The inhibition is partial (hyperbolic) in the presence of excess magnesium (free concentration about 5 mM) (Fig 7.1); i.e. the half-saturation constant ($K_{0.5}$) for glucose was increased but at all glucose concentrations the activity was greater than zero at the highest palmitoyl-CoA concentrations tested, and the measured critical micelle concentration under the same conditions was higher than the apparent K_i, which was quite low. Perturbation of micelle formation did not affect the type of inhibition and the K_i for palmitoyl-CoA was always several times below the critical micelle concentration,[15] indicating that the incompleteness of the inhibition was not due to the formation of micelles.

Apparent inhibition constants were calculated to be 1.8 μM for palmitoyl-CoA and 0.77 μM for oleoyl-CoA, apparently within the range of intrahepatic concentrations.[15] The true K_i estimated at zero glucose and ATP concentrations is even lower, 0.60 μM. Palmitoyl-CoA had a similar partial competitive K_i with respect to both glucose and MgATP. In addition, the cooperativity with glucose was independent of the acyl-CoA concentrations ($h = 1.5 \pm 0.05$) as can be seen in Figure 7.1, in contrast to what happens with substrate analogs such as mannose or *N*-acetylglucosamine,[16,17] as we saw in Chapter 6.

Taken together, these data on the specificity, kinetics, and character of the inhibition of hexokinase D by long chain acyl-CoAs, which are unaffected by perturbation of micelle formation and the partition of amphiphiles, suggest that the inhibition is site-specific and unrelated either to the formation of detergent micelles or to nonspecific hydrophobic binding of amphiphiles to hexokinase D.

Mechanism of action

The basic interpretation of the inhibition by palmitoyl-CoA (and oleoyl-CoA) is that binding at an allosteric site elicits a structural change in the enzyme that shifts all forms of it to a new set with different activities. This implies that they are all still active, with lower affinity for glucose and MgATP, but that there is no change in the relationships between the rate constants that give positive cooperativity with a Hill coefficient of about 1.5.

As we saw in Chapter 6, there are three potential mechanisms for explaining the kinetic cooperativity of hexokinase D with respect to glucose, and the observations with palmitoyl-CoA can be explained in terms of the ligand-induced slow-transition mechanism (the LIST mechanism in Neet's terminology). Thus Neet and colleagues demonstrated by model simulations that the kinetic behavior of hexokinase D in the presence of palmitoyl-CoA, which does not alter the cooperativity with glucose, was consistent with this mechanism.[18] Changes in the *on/off* rate constants for glucose binding by factors of less than 10 resulted in shifts of more than 20-fold in the $K_{0.5}$ without significant change in the Hill coefficient. In contrast, they were unable to show comparable effects with the mnemonical mechanism: the *on/off* rates appear to affect the Hill coefficient more in this mechanism because of the limitations due to having only a single catalytic cycle.

The inhibition observed with palmitoyl-CoA itself was cooperative, with a Hill coefficient for the inhibition of about 1.7 regardless of whether the varied substrate was glucose or ATP, and incomplete, i.e. the rate did not approach zero at saturation with inhibitor.[15] However, the reasons for this complexity have not been determined. In an attempt to clarify matters Neet and colleagues studied the binding of palmitoyl-CoA,[19] but the results were dominated by a slow, irreversible binding of a large amount of palmitoyl-CoA and did not relate to the specific allosteric binding required by the reversible inhibition. In my opinion, with which Neet concurs (personal communication), the behavior probably reflects the complex kinetics involving the slow transition, as this interpretation requires the fewest additional assumptions, requiring only a single palmitoyl-CoA site.

The capacity of *N*-acetylglucosamine to decrease the cooperativity of hexokinase D with respect to glucose,[17] which palmitoyl-CoA lacks,[14] emphasizes the difference in kinetic effects that are obtained from binding to all forms of the enzyme (palmitoyl-CoA) or preferentially to only some forms (*N*-acetylglucosamine).

Effect of Mg^{2+} on the inhibition

At this point I want to draw attention again to the importance of correctly controlling the Mg^{2+} and ATP concentrations in the assay medium. Neet's group studied the kinetics of acyl-CoA inhibition under two sets of conditions, one with equimolar concentrations of $MgCl_2$ and ATP (low Mg^{2+}), and the other with a constant 5 mM excess of $MgCl_2$ over ATP (high Mg^{2+}). As discussed in Chapter 3, these conditions imply quite different distributions of ionic species, especially free Mg^{2+} and free ATP anions (see Figure 3.11). Although the results obtained were similar in the two conditions in relation to the competitive inhibition with respect to both substrates and the lack of effect on glucose cooperativity, they nevertheless differed in an important respect. With equimolar $MgCl_2$ and ATP (low Mg^{2+}) the inhibition by acyl-CoA (apart from arachidoyl-CoA) was complete, i.e. the rate approached zero at saturation, but the inhibition could be made incomplete by adding Mg^{2+}.[14] The mechanism of this effect of Mg^{2+} is unclear, but it appears that it binds to acyl-CoA (as the critical micelle concentration is affected by Mg^{2+}), as may be seen in Table 7.1. A possibility suggested by the authors is that free acyl-CoA and the complex between acyl-CoA with Mg^{2+} could both bind to an allosteric site,[15] inducing somewhat different conformational changes, so that binding of free acyl-CoA would produce a completely inactive enzyme whereas binding of the complex would produce an enzyme with some residual activity.

In vivo, total Mg^{2+} and ATP are not equimolar, as the free Mg^{2+} concentration is buffered in the range 0.5-1 mM, i.e. to a first approximation we can say that the cell uses the experimental design of maintaining total Mg^{2+} in constant excess of total ATP. Consequently any physiological effect of acyl-CoA is likely to involve only partial inhibition of hexokinase D.

Physiological role

Assessment of any physiological role of the inhibition of hexokinase D activity by fatty acyl-CoAs with certainty is not possible at this stage. However, the idea of some control over the uptake and storage of blood glucose by the major product of fat metabolism through the substrate cycle between glucose and glucose 6-phosphate is very appealing. One may speculate, therefore, as Neet's group did,[15] that physiological concentrations of palmitoyl-CoA perform a regulatory role by shifting this cycle toward release of glucose during starvation or diabetes when lipolysis and lipid metabolism are increasing. In conditions that stimulate lipolysis, such as low insulin and high glucagon concentrations, substantial amounts of fatty acids arrive in the liver and produce acyl-CoA there (Fig. 7.2). The liver contains about 110 nmol of long-chain fatty acids per gram of protein in fed rats, and about 250 nmol in fasted rats,[20] corresponding roughly to concentrations of 45 µM and 100 µM respectively. The free concentrations in the cytosol may be considerably lower as a consequence of subcellular compartmentation and binding to other proteins, in particular to an acyl-CoA-binding protein. However, as the inhibition constants are very small (Table 7.1), even concentrations ten times lower would be sufficient for some inhibition. The case of stearoyl-CoA is particularly interesting, as its concentration varies about six-fold from fed to fasted animals (4.8 to 29 nmol per gram of protein, or about 2 to 12 µM). As its apparent inhibition constant is only 0.32 µM (Table 7.1), substantial inhibition is possible.

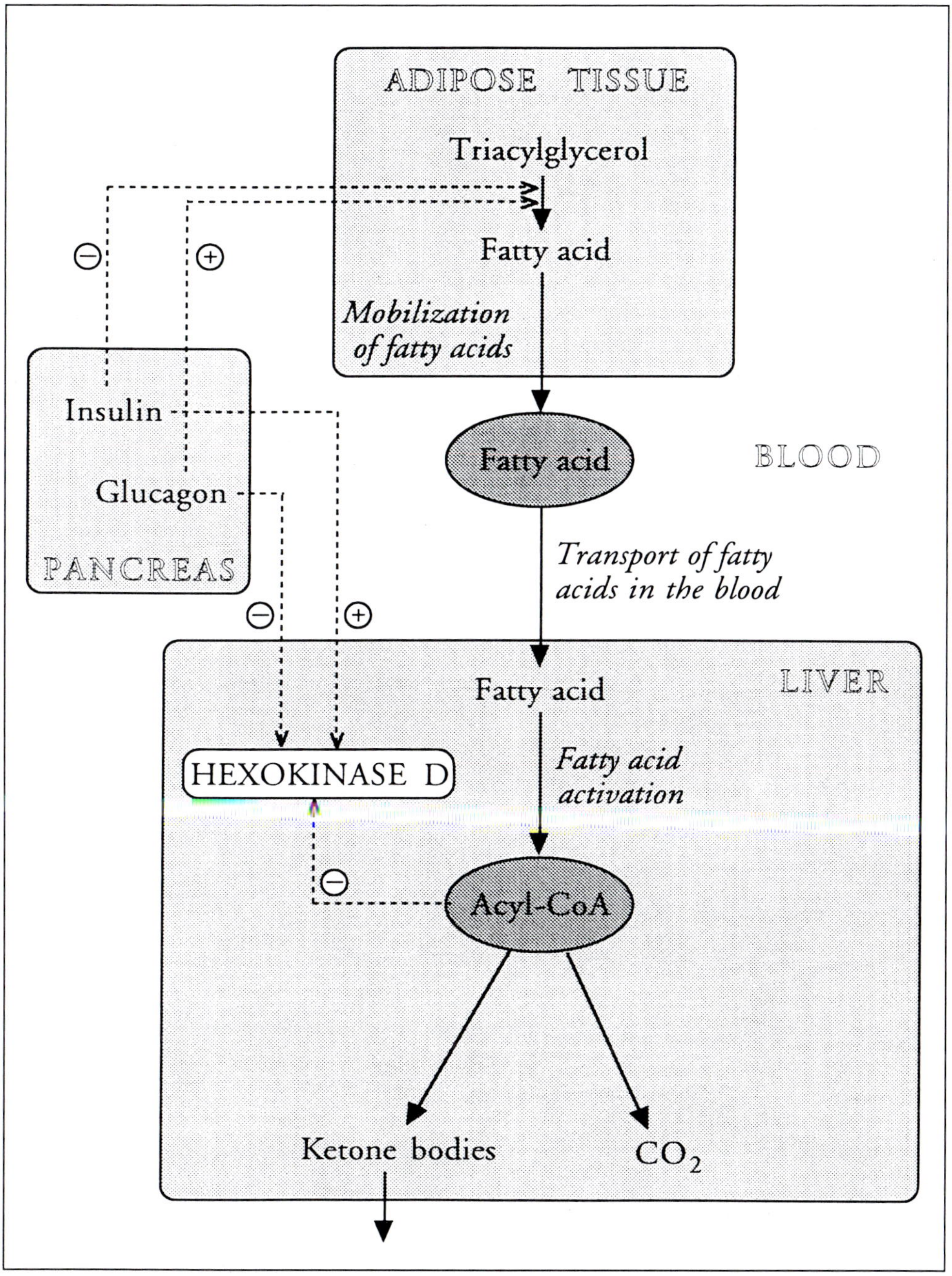

Fig. 7.2. Physiological interrelationships between fatty acids and hexokinase D activity.

I shall discuss this point further below in relation to the regulatory protein.

REGULATION OF HEXOKINASE D BY A REGULATORY PROTEIN

In this section two reviews from Van Schaftingen's group are of particular relevance.[21,22]

SOME INTRIGUING FACTS

The fructose effect

It has long been well known that fructose stimulates the utilization of glucose by the liver. The rate of glycogen formation increases in the presence of fructose, so there is a synergism between glucose and fructose. For example, Seglen observed in an experiment done with isolated parenchymal cells that the rate of glycogen synthesis at 5 mM fructose and 20 mM glucose was more than ten times as high as the sum of the glycogenic rates obtained with the two sugars separately.[23] Other metabolites, such as lactate[23] and glyceraldehyde,[24] had similar but much less striking effects in liver. For glyceraldehyde the effect was very minor, and the lack of inhibition was especially striking.[24] This raised doubts about the participation of the hexokinase reaction in glycogen synthesis.

Various authors had proposed that the stimulation by fructose resulted from a decrease in the breakdown of newly deposited glycogen brought about by inhibition of glycogen phosphorylase by millimolar concentrations of fructose 1-phosphate.[25,26] However, this appeared not to be the whole explanation, as fructose at physiological concentrations appeared to stimulate (4-fold at 2.5 mM) the incorporation of [U-^{14}C]glucose into metabolic products, principally CO_2, lactate and pyruvate,[27] suggesting a possible stimulatory effect at the level of glucose phosphorylation. The concentrations of fructose in blood in the hepatic portal vein of rats consuming a meal of fructose are in the range of 1.5-2.5 µmol/ml of blood.[28] As a matter of fact, the same authors showed that formation of ^3H_2O from 2-^{3}H-glucose, a measure of the hexokinase D reaction in hepatocytes, increases in isolated hepatocytes in response to the addition of 2.5 mM fructose. As ^{3}H at C2 of glucose is lost during isomerization catalyzed by glucose 6-phosphate isomerase, and as glucose 6-phosphate is always in equilibrium with fructose 6-phosphate because of the high activity of this enzyme, the rate of detritiation of 2-^{3}H-glucose depends both on the rate of glucose phosphorylation and on the extent of detritiation of the glucose 6-phosphate pool.[29]

Thus the effect of fructose on the glucose-phosphorylating activity of hepatocytes appeared intriguing, and raised the question of whether it could be explained in terms of the known effects of fructose on hexokinase D, the main hexokinase isoenzyme in the hepatocyte. As we have seen (Chapter 6), fructose does influence hexokinase D kinetics, being very effective at suppressing the cooperativity. It is potentially capable, therefore, of stimulating hexokinase D activity at low glucose concentrations. However, although the fructose concentration needed for this is low compared with the K_m it is still much higher than the range of physiological concentrations, and any such direct effect by stimulating hexokinase D activity must be much smaller than the effects seen in hepatocytes. Consequently the mechanism by which fructose could stimulate glucose utilization remained far from clear.

A much higher $K_{0.5}$ of hexokinase D for glucose in hepatocytes than in the pure state

As we have seen in Chapter 3, purified hexokinase D has a $K_{0.5}$ of about 5 mM at pH 8.0 and of about 8.0-9.0 mM at the more physiological pH of 7.3, with a similar Hill coefficient of about 1.5 in the two cases. The presence of K^+ or Na^+ in the assay medium does not modify the Hill coefficient and both ions produce similar slight decreases of the V_{max} and the $K_{0.5}$.[30] In isolated hepatocytes, on the other hand, Hers's group reported that the $K_{0.5}$ of glucose phosphorylation measured radiochemically through the release of ^3H from 2-^3H-glucose, is higher than 9 mM, and the value depends on the ionic composition.[31] The saturation curve was markedly more sigmoidal in a Na^+ medium (hepatocytes incubated in Krebs-Ringer bicarbonate buffer) with a Hill coefficient of about 1.8 and $K_{0.5}$ about 20 mM, than in a K^+ medium, in which the corresponding values were 1.3 and 12 mM respectively. At that time there was no explanation for these results, but this enhancement of glucose phosphorylation induced by K^+ in isolated hepatocytes agreed well with the great enhancement of glycogen synthesis that the same group had previously reported to occur in hepatocytes incubated in a K^+ medium.[32]

This higher $K_{0.5}$ and the effect of different ions disappeared when the experiment was done with a hepatocyte extract; $K_{0.5}$ was then close to 9 mM and the Hill coefficient was 1.4-1.5, similar to the parameters obtained with the purified enzyme at similar pH,[30] suggesting that some structural requirement was lost during preparation of the extract and the concomitant dilution.

Discovery of the Regulatory Protein

In an attempt to understand these observations at a molecular level, Emile Van Schaftingen, a former member of Hers's group, reexamined the fructose and K^+ effects. He confirmed[33] previous results and made the important discovery that rat liver contains a regulatory protein that is able in the presence of fructose 6-phosphate to bind to hexokinase D and to inhibit its activity.[7] Fructose 1-phosphate is able to act as a positive effector, counteracting the inhibition exerted by

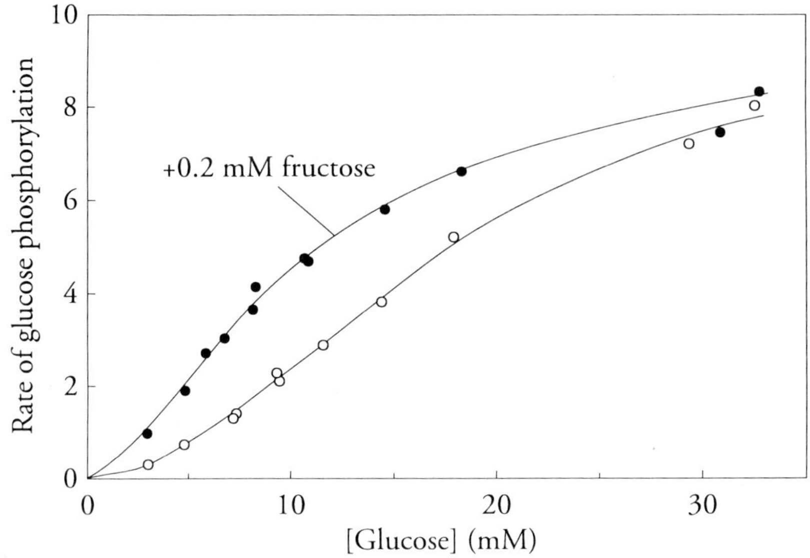

Fig. 7.3. Stimulation by fructose of glucose phosphorylation in isolated hepatocytes. Hepatocytes isolated from an overnight-fasted rat were incubated at different glucose concentrations in the presence or absence of 0.2 mM fructose. Glucose phosphorylation was followed by measuring detritiation of [2-^{3}H]glucose. Redrawn from Eur J Biochem 1989; 179:173-177.

fructose 6-phosphate, which inhibits indirectly. This interplay between the regulatory protein and the two phosphate esters allows the stimulatory effect of fructose to be explained. Neither fructose 6-phosphate not fructose 1-phosphate are effectors of purified hexokinase D, at least not at these low concentrations; they need the presence of a protein of about 60 kDa that Van Schaftingen has called the regulatory protein of hexokinase D.

Working in isolated rat hepatocytes, he showed that 0.2 mM fructose stimulates 2-4-fold the rate of glucose phosphorylation at 5 mM glucose, but not at 40 mM, thus increasing the apparent affinity of the glucose phosphorylating system. (He used this small concentration of fructose to avoid any glycogenic effect and inhibition of glucose detritiation, as fructose 1-phosphate appears to inhibit phosphoglucoisomerase.) When hepatocytes isolated from a rat fasted overnight were incubated in the presence and absence of 0.2 mM fructose, the effect of fructose was to decrease the $K_{0.5}$ for glucose from 15-20 mM to 9-10 mM, phosphorylation of glucose being followed as detritiation of 2-^{3}H-glucose (Fig. 7.3). In the presence of fructose the hepatocyte thus shows a similar $K_{0.5}$ for glucose to that of the extract[31] or the purified enzyme at similar pH.[30] The effect was almost maximal 5 minutes after addition of fructose and it disappeared when the cells were washed free of it, indicating that the increase in hexokinase D activity in the presence of fructose was not due to synthesis of a new protein but to a more rapid process, such as an allosteric effect or a covalent modification.[33] D-Glyceraldehyde also had a stimulatory effect, but at tenfold higher concentrations, whereas dihydroxyacetone did not.[33]

By contrast, oleate did not affect the phosphorylation of glucose in hepatocytes fasted overnight, even in the presence of fructose, although it stimulated the formation of ketone bodies severalfold, indicating that it was converted to its acyl-CoA derivative. The authors interpret this result to suggest that long chain fatty acyl-CoAs that inhibit the purified enzyme would not be important in the hepatocyte.[33] However, as metabolic regulation is often more complex than we expect, it is possible that to be able to detect inhibition of the

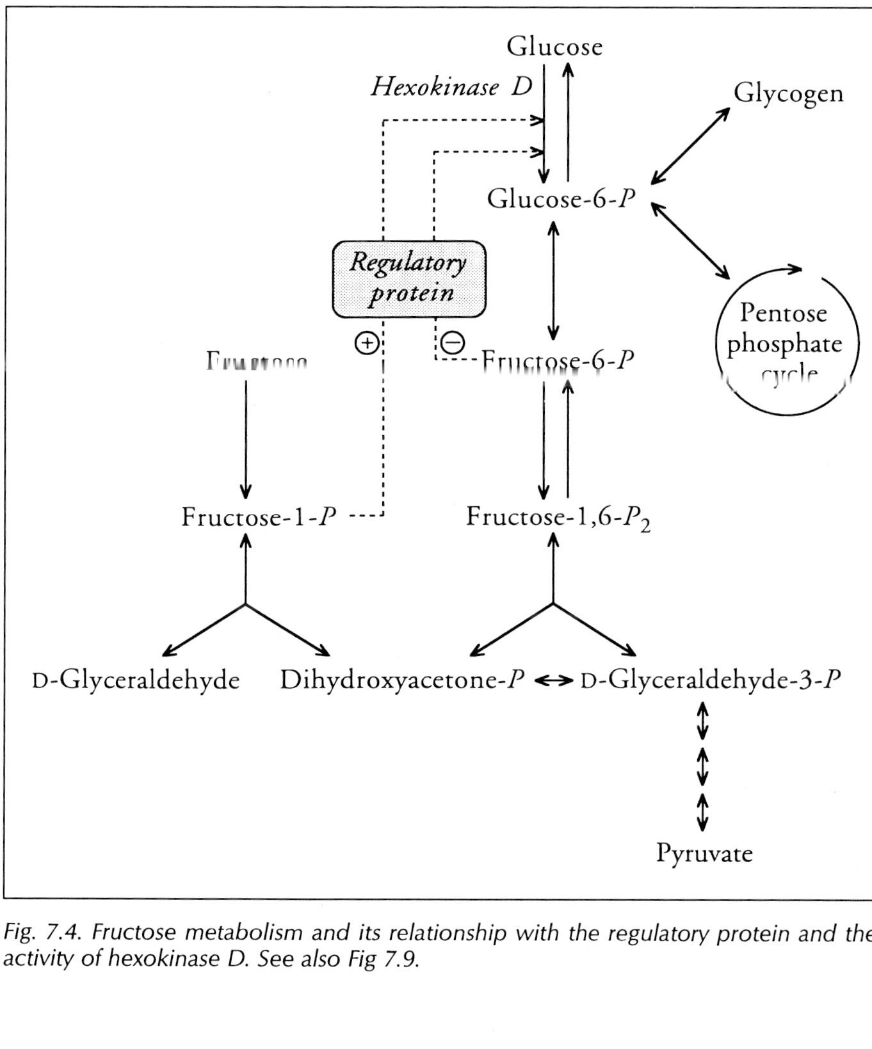

Fig. 7.4. Fructose metabolism and its relationship with the regulatory protein and the activity of hexokinase D. See also Fig 7.9.

hexokinase D reaction by long chain fatty acyl-CoAs in vivo more accentuated conditions of lipolysis could be required. One must remember that hexokinase D is not the only candidate for inhibition in vivo by these compounds and so it would be interesting to know if an effect is observed with hepatocytes from diabetic rats or from rats fasted longer than just overnight. This is discussed further at the end of the chapter.

The fact that glucose detritiation is stimulated by fructose, sorbitol and glyceraldehyde but not by glycerol or dihydroxyacetone, and that fructose itself is more effective than glyceraldehyde, suggested that fructose does not act directly but through a metabolite, most probably fructose 1-phosphate. As can be seen in Figure 7.4 fructose 1-phosphate is a normal intermediate of fructose and sorbitol metabolism and it can also be formed from glyceraldehyde by condensation with dihydroxyacetone phosphate.

As fructose 1-phosphate does not affect the kinetics of the pure enzyme, whereas fructose increases the affinity of hexokinase D for glucose in the hepatocyte, it is sensible to postulate, as Van Schaftingen did, that it causes a competitive inhibitor present in the hepatocyte to be

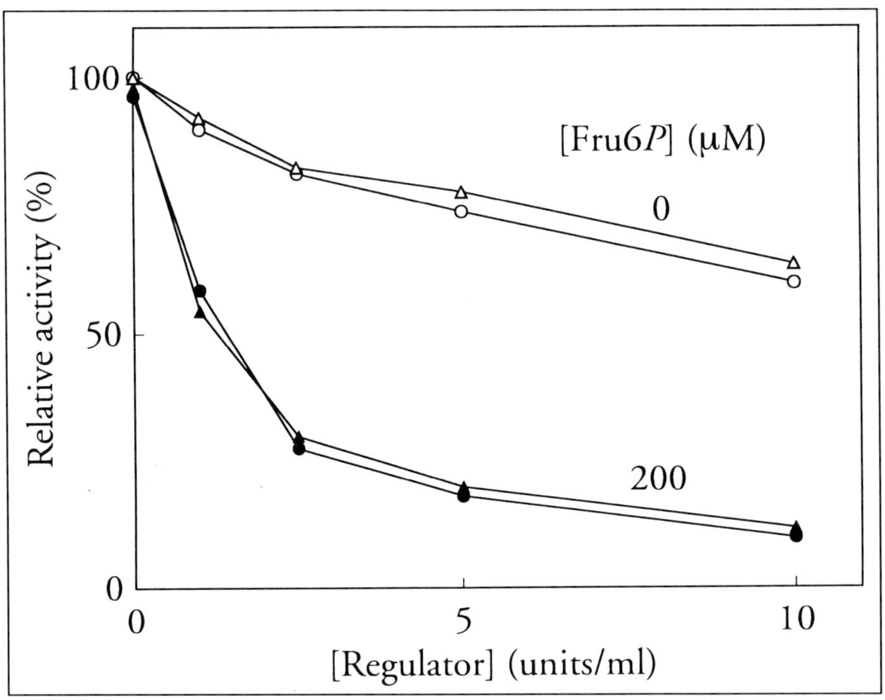

Fig. 7.5. Inhibition of hexokinase D by the regulatory protein in the presence and absence of 200 μM fructose 6-phosphate. The circles and triangles represent duplicate experiments. Reproduced with permission from Eur J Biochem 1991; 200:553-561.

eliminated or counteracted. The existence of such an inhibitor in the hepatocytes would explain the higher $K_{0.5}$ for glucose in the cell compared with the value for the purified enzyme and the fact that the stimulatory effect of fructose is lost at high glucose concentrations. In addition, although fructose 1-phosphate does not stimulate purified hexokinase D it does stimulate glucose phosphorylation in liver extracts. These two observations put together suggested that the regulation of hexokinase D by fructose 1-phosphate was due to a factor lost early in the purification. (Hexokinase D does not show variations in kinetic behavior during purification once it has been separated from the other isoenzymes,[34] as we saw in Chapter 3.) Searching for this inhibitor, Van Schaftingen showed that anion-exchange chromatography of liver extracts separated hexokinase D from a protein that inhibited it, and that the inhibition was abolished by the presence of fructose 1-phosphate.[7]

The inhibitory protein was purified and shown to inhibit hexokinase D (Fig. 7.5).[7,35] The inhibition was greatly reinforced by low concentrations of fructose 6-phosphate, with a half-maximal effect at a concentration of about 10 μM, but not by glucose 6-phosphate. (In vivo, however,

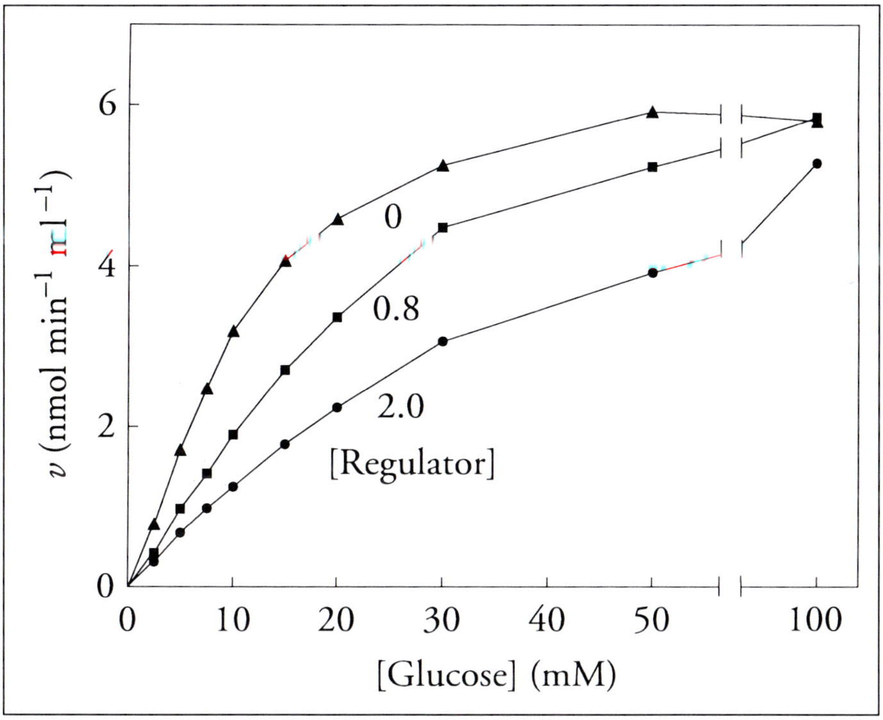

Fig. 7.6. Effect of the regulatory protein on the affinity for glucose of hexokinase D. The experiment was done in the presence of 200 μM fructose 6-phosphate and the concentrations of glucose and regulator indicated. Reproduced with permission from Eur J Biochem 1989; 179:179-184.

glucose 6-phosphate is in equilibrium with fructose 6-phosphate, and so an indirect inhibitory effect is to be expected). The inhibitory effect of the regulatory protein is observed in conditions where there is reasonable certainty that there is no contamination by glucose 6-phosphate isomerase or fructose 6-phosphate,[35] and so the small residual effect seen in the absence of fructose 6-phosphate must be due to the inhibitory protein by itself.

In the presence of fructose 6-phosphate the inhibition of hexokinase D by the regulatory protein is competitive with respect to glucose (Fig. 7.6).[7,36] The $K_{0.5}$ for glucose increases linearly as a function of the concentration of the inhibitory mixture (regulatory protein together with fructose 6-phosphate), indicating a complete type of inhibition, and there is a small decrease in the sigmoidicity of the saturation curve, the Hill coefficient decreasing from 1.58 to 1.32.[36] Fructose 1-phosphate relieved the inhibitory effect of the regulatory protein and antagonized the effect of fructose 6-phosphate (Fig. 7.7) in a competitive manner.[7] All these experiments were done with equimolar total concentrations of Mg^{2+} and ATP, which as discussed above in relation to inhibition

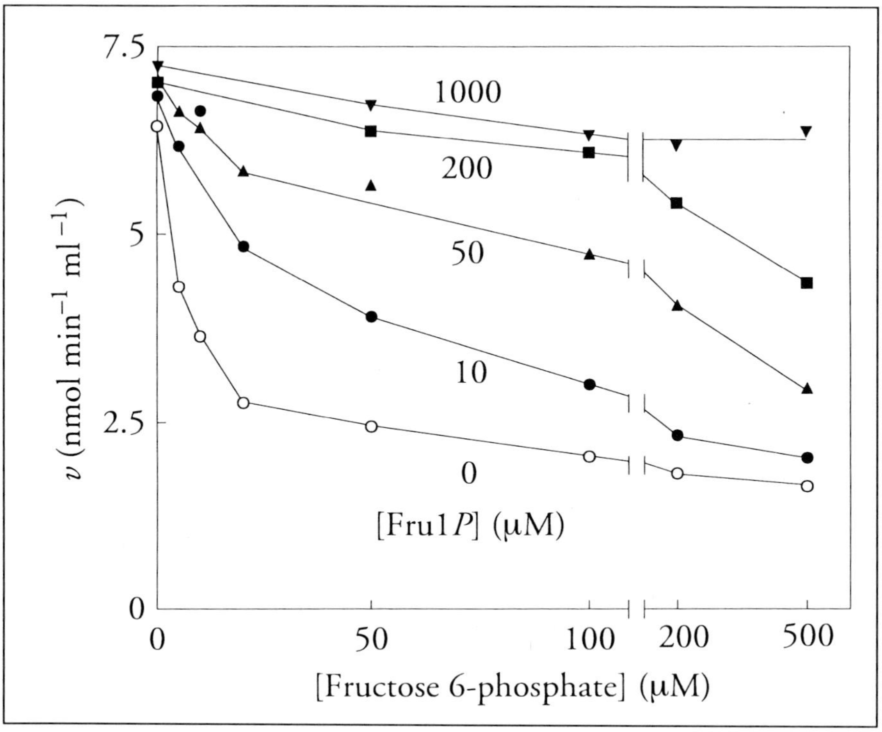

Fig. 7.7. Antagonism between fructose 1-phosphate and fructose 6-phosphate in the regulation of hexokinase D. Fructose 1-phosphate reverses the inhibition exerted by fructose 6-phosphate in the presence of the regulatory protein. Reproduced with permission from Eur J Biochem 1989; 179:179-184.

by acyl-CoA, and also in Chapter 3, complicates their interpretation somewhat. It would be interesting to know whether the changed inhibitory behavior of acyl-CoA in the presence of excess Mg^{2+} is paralleled by the regulatory protein.

The fact that the inhibitory effect of the regulator sensitive to the fructose phosphates was observed in three different types of assay, measuring either the production of glucose 6-phosphate (including a radiochemical assay), or that of ADP, rules out an assay artifact and argues for a true inhibition of hexokinase D.

The presence of the regulatory protein now explains the fact that the apparent affinity of hexokinase D for glucose is higher in isolated hepatocytes than with the isolated enzyme.[31] The effect of fructose in liver is thus due to the fact that it is the substrate of fructokinase, which phosphorylates fructose at C1 and produces fructose 1-phosphate, thereby activating glucose phosphorylation by decreasing the inhibition by the regulatory protein. Any metabolite that can cause production of fructose 1-phosphate ought to have the same effect. On the other hand, an increase in the concentration of fructose 6-phosphate will increase the inhibition of hexokinase D by the regulatory protein.

From the amount of regulator recovered after anion-exchange chromatography, it can be calculated that the liver contains at least 25 units of regulator per gram, a unit of regulator being defined by Van Schaftingen as the amount causing half-maximal inhibition of hexokinase D under the conditions used in the pyruvate kinase/lactate dehydrogenase coupled assay, with 5 mM glucose and 200 μM fructose 6-phosphate.[7]

As the concentration of fructose 6-phosphate in the liver is usually above 10 μM,[37] the amount of regulatory protein is sufficient to inhibit hexokinase D to a large extent. The inhibitory effect of the regulatory protein provides, therefore, an explanation for the observation that the apparent affinity of hexokinase D for glucose is lower in intact hepatocytes than expected from assays performed with purified enzyme or cell-free extracts, under conditions that do not allow hexose 6-phosphates to accumulate. The fact that fructose, together with other compounds giving rise to fructose 1-phosphate, restore the affinity to a normal level is in accordance with this interpretation. On the other hand it does not explain why the observed Hill coefficient was much higher.

Assay, purification and characterization of the regulatory protein

The rat liver regulatory protein has been purified to apparent homogeneity by two slightly different procedures.[36,38] It corresponds to a single polypeptide of 62 kDa by SDS-PAGE, 60 kDa by gel filtration and 68 kDa by centrifugation in an isokinetic sucrose gradient. It is assayed by its ability to inhibit rat liver hexokinase D.[7]

GAGGGTCCACAGTGTGGGACCATGCCAGGCACCAAACGATATCAGCATGTGATCGAGACCCCTGAGCCTGGTGAA 75
Met Pro Gly Thr Lys Arg Tyr Gln His Val Ile Glu Thr Pro Glu Pro Gly Glu

TGGGAGTTGTCAGGGTATGAAGCGGCTGTGCCAATCACAGAGAAATCCAACCCACTGACCCGAAACCTGGACAAA 150
Trp Glu Leu Ser Gly Tyr Glu Ala Ala Val Pro Ile Thr Glu Lys Ser Asn Pro Leu Thr Arg Asn Leu Asp Lys

GCAGATGCAGAGAAAATTGTCAAACTGCTGGGGCAGTGTGATGCTGAGATATTCCAGGAGGAGGGGCAGATTGTG 225
Ala Asp Ala Glu Lys Ile Val Lys Leu Leu Gly Gln Cys Asp Ala Glu Ile Phe Gln Glu Glu Gly Gln Ile Val

CCCACCTACCAGCGACTATACAGCGAATCAGTTCTGACCACCATGTTGCAAGTGGCTGGAAAAGTCCAGGAAGTT 300
Pro Thr Tyr Gln Arg Leu Tyr Ser Glu Ser Val Leu Thr Thr Met Leu Gln Val Ala Gly Lys Val Gln Glu Val

CTGAAGGAGCCAGATGGGGGTCTGGTAGTGCTGAGTGGAGGGGGAACCTCTGGTCGTATGGCATTTCTCATGTCT 375
Leu Lys Glu Pro Asp Gly Gly Leu Val Val Leu Ser Gly Gly Gly Thr Ser Gly Arg Met Ala Phe Leu Met Ser

GTGTCTTTCAACCAGCTGATGAAAGGCCTGGGACAAAAGCCTCTTTACACCTACCTCATTGCAGGAGGTGACAGG 450
Val Ser Phe Asn Gln Leu Met Lys Gly Leu Gly Gln Lys Pro Leu Tyr Thr Tyr Leu Ile Ala Gly Gly Asp Arg

TCTGTTGTGGCCTCTCGTGAACAGACAGAAGATAGCGCCCTACACGGGATCGAGGAGCTGAAGAAGGTGGCTGCT 525
Ser Val Val Ala Ser Arg Glu Gln Thr Glu Asp Ser Ala Leu His Gly Ile Glu Glu Leu Lys Lys Val Ala Ala

GGGAAGAAGAGAGTGGTCGTCATAGGCATCTCTGTGGGACTCTCTGCGCCCTTTGTGGCAGGTCAGATGGACTAC 600
Gly Lys Lys Arg Val Val Val Ile Gly Ile Ser Val Gly Leu Ser Ala Pro Phe Val Ala Gly Gln Met Asp Tyr

TGCATGGATAACACAGCCGTCTTCTTGCCGGTTCTGGTTGGCTTCAATCCAGTGAGCATGGCCAGAAATGACCCC 675
Cys Met Asp Asn Thr Ala Val Phe Leu Pro Val Leu Val Gly Phe Asn Pro Val Ser Met Ala Arg Asn Asp Pro

ATTGAAGACTGGAGATCAACATTCCGGCAAGTGGCAGAGCGGATGCAAAAGATGCAGGAGAAACAGGAAGCTTTT 750
Ile Glu Asp Trp Arg Ser Thr Phe Arg Gln Val Ala Glu Arg Met Gln Lys Met Gln Glu Lys Gln Glu Ala Phe

GTGCTCAATCCTGCCATCGGGCCCGAGGGGCTCAGCGGCTCTTCCCGAATGAAAGGTGGAGGTGCCACCAAGATT 825
Val Leu Asn Pro Ala Ile Gly Pro Glu Gly Leu Ser Gly Ser Ser Arg Met Lys Gly Gly Gly Ala Thr Lys Ile

CTACTGGAAACCCTGCTACTAGCAGCCCATAAGACTGTGGACCAGGGTGTTGTGTCCTCTCAAAGATGCCTTCTG 900
Leu Leu Glu Thr Leu Leu Leu Ala Ala His Lys Thr Val Asp Gln Gly Val Val Ser Ser Gln Arg Cys Leu Leu

GAAATCCTGAGGACATTTGAGCGGGCTCATCAGGTGACCTACAGTCAAAGTTCCAAAATTGCCACGCTGATGAAA 975
Glu Ile Leu Arg Thr Phe Glu Arg Ala His Gln Val Thr Tyr Ser Gln Ser Ser Lys Ile Ala Thr Leu Met Lys

CAAGTCGGCATCAGCCTGGAGAAGAAAGGCCGAGTGCACTTGGTTGGCTGGCAGACTCTCGGCATCATTGCCATT 1050
Gln Val Gly Ile Ser Leu Glu Lys Lys Gly Arg Val His Leu Val Gly Trp Gln Thr Leu Gly Ile Ile Ala Ile

ATGGACGGAGTAGAGTGCATCCACACTTTTGGTGCTGATTTCCAAGATATCCGTGGCTTTCTTATTGGTGACCAC 1125
Met Asp Gly Val Glu Cys Ile His Thr Phe Gly Ala Asp Phe Gln Asp Ile Arg Gly Phe Leu Ile Gly Asp His

AGTGACATGTTTAACCAGAAGGATGAACTCACCAACCAGGGTCCCCAGTTCACCTTCTCCCAGGATGACTTCCTG 1200
Ser Asp Met Phe Asn Gln Lys Asp Glu Leu Thr Asn Gln Gly Pro Gln Phe Thr Phe Ser Gln Asp Asp Phe Leu

ACTTCCATCCTGCCATCCCTCACGGAGACTGACACCGTGGTCTTCATTTTTACCCTGGATGATAACCTCACAGAA 1275
Thr Ser Ile Leu Pro Ser Leu Thr Glu Thr Asp Thr Val Val Phe Ile Phe Thr Leu Asp Asp Asn Leu Thr Glu

GTACAGGCCCTGGCAGAAAGAGTGAGAGAGAAGTGCCAGAACATCCAGGCCCTGGTGCACAGCACTGTGGGGCAG 1350
Val Gln Ala Leu Ala Glu Arg Val Arg Glu *Lys Cys Gln Asn Ile Gln Ala Leu Val His Ser Thr Val Gly Gln*

TCCTTGCCGGCCCCTCTAAAGAAACTCTTTCCCTCACTCATCAGTATCACGTGGCCACTTCTTTTCTTCGATTAT 1425
Ser Leu Pro Ala Pro Leu Lys Lys Leu Phe Pro Ser Leu Ile Ser Ile Thr Trp Pro Leu Leu Phe Phe Asp Tyr

GAAGGGACCTATGTTCAGAAGTTCCAGCGTGAGTTAAGCACCAAGTGGGTGTTGAATACAGTGAGTACTGGGGCC 1500
Glu Gly Thr Tyr Val Gln Lys Phe Gln Arg Glu Leu Ser Thr Lys Trp Val Leu Asn Thr Val Ser Thr Gly Ala

CATGTACTGCTGGGGAAGATCCTACAGAACCACATGCTGGACCTCCGCATCGCCAACTCCAAGCTCTTCTGGAGG 1575
His Val Leu Leu Gly Lys Ile Leu Gln Asn His Met Leu Asp Leu Arg Ile Ala Asn Ser Lys Leu Phe Trp Arg

GCGCTGGCCATGTTGCAGAGGTTCTCTGGACAGTCCAAGGCTCGCTGCATTGAGAGCCTCCTTCAAGCAATCCAC 1650
Ala Leu Ala Met Leu Gln Arg Phe Ser Gly Gln Ser Lys Ala Arg Cys Ile Glu Ser Leu Leu Gln Ala Ile His

TTTCCTCAACCACTGTCGGATGATGTCCGCGCCGCTCCCATCTCCTGCCACGTCCAGGTTGCCCACGAGAAGGAA 1725
Phe Pro Gln Pro Leu Ser Asp Asp Val Arg Ala Ala Pro Ile Ser Cys His Val Gln Val Ala His Glu Lys Glu

AAGGTGATCCCCACAGCCTTGCTGAGCCTCCTACTCCGGTGCTCCATCTCTGAGGCTAAGGCACGCCTGTCTGCA 1800
Lys Val Ile Pro Thr Ala Leu Leu Ser Leu Leu Leu Arg Cys Ser Ile Ser Glu Ala Lys Ala Arg Leu Ser Ala

GCTTCTTCAGTCTGTGAGGTTGTTAGGAGCGCCCTCTCTGGGCCGGGTCAGAAGCGCAGCACGCAAGCCCTTGAA 1875
Ala Ser Ser Val Cys Glu Val Val Arg Ser Ala Leu Ser Gly Pro Gly Gln Lys Arg Ser Thr Gln Ala Leu Glu

GACCCTCCCGCCTGTGGGACCCTGAATTGATATTTCTAGAACCATGGAGGGGCAGAGTCTCCGTCCACTTCCAAG 1950
Asp Pro Pro Ala Cys Gly Thr Leu Asn

GGGACATGTGCCAGCAGTACACGCTGTGGGAAGAACTCAGTTTCGGGTGGGTGGGGCCTAACTGCCCAGAATTGG 2025

GGAAGAGCCCTGTTCTCAACCGGATTATTTCCATTTTTACTGGTGTCTTCTGAACTCAGAAATAAAACTAAATGT 2100

CTTGTTTTGGAAAAAAAAAAAAAAAAAAAAAAAAAAAAAAAAAAAAAAAAAA

Fig. 7.8. Sequence of the cDNA corresponding to the regulatory protein, together with the deduced amino acid sequence. Note that the italicized amino acids (from Lys at base position 1305 to Leu at base position 1441) have been corrected from those given both in the original article and its erratum,[40] both of which showed the Lys-Cys moved from the beginning to the end of this region. The amino-acid sequence deposited in the EMBL data bank did not contain this error (E. Van Schaftingen, personal communication).

Because of interference by the hexokinase isoenzymes this assay cannot be used with crude extracts, and consequently the regulatory protein has to be separated from the hexokinases, which can be achieved by partial purification on DEAE-sepharose. A procedure for processing samples of tissues of 0.5-1 g has been described.[39]

The gene has been cloned and sequenced. Partial peptide sequences of the rat liver protein were obtained and used to clone the corresponding cDNA,[40] which encodes a polypeptide of 628 aminoacids (not 568 as originally reported).[22,40] This sequence is shown in Figure 7.8. Northern blots indicate the presence in rat liver of a major RNA species of 2.35 kb and also a minor band of 3.9 kb, which is possibly an immature transcript.[40]

As we mentioned in Chapter 3, hexokinase D is half the size of the other mammalian hexokinases. It was very tempting when the existence of the regulatory protein was first reported to speculate that it might correspond to the second (*N*-terminal) half of the hexokinase gene that had evolved to provide an allosteric site for fructose 6-phosphate rather than for glucose 6-phosphate. This hypothesis, although very attractive, can now be ruled out completely, for several reasons:

1. The structural organization of the hexokinases is now realized to be more complex than was thought initially. The allosteric site for glucose 6-phosphate, if it exists, appears to be very near the catalytic site; the same half of the molecule of hexokinase A (*C*-terminal half) appears to provide both activities, i.e. catalysis and the high sensitivity to inhibition by glucose 6-phosphate,[41-43] as mentioned in Chapter 2.
2. Specificity studies of the effectors of the regulatory protein show that analogs of the open-chain configuration of fructose 6-phosphate, such as sorbitol 6-phosphate, are powerful inhibitors, indicating that it is the open-chain configuration that is recognized by the regulatory protein.[35] This is not the case of the hexokinases, which are not inhibited by sorbitol 6-phosphate.[44] Consequently it is more likely that the regulatory protein is derived from an enzyme with affinity for the linear form of fructose 6-phosphate or fructose 1-phosphate, such as glucose-phosphate isomerase, aldolase, transaldolase and transketolase.[35]
3. Finally, the polypeptide encoded by the cDNA corresponding to the regulatory protein shows no significant evidence of homology with any known protein sequences.[22,40]

Mechanism of action

The regulatory protein appears to form a complex with hexokinase D, as the molecular mass of the enzyme goes from a value of about

55 kDa, in agreement with previous work,[45] to more than 100 kDa when it is in the presence of the regulatory protein and fructose 6-phosphate, in sucrose gradient sedimentation.[38] This increase in molecular mass depends on the presence of fructose 6-phosphate, and is not observed in its absence or if fructose 1-phosphate is also present.[38] However, the fact that the regulatory protein can inhibit hexokinase D in the absence of fructose 6-phosphate (about 7% of the maximal inhibition) means that a fraction of hexokinase D must be complexed in those conditions. In principle this should be seen as a small broadening of the peak in the sedimentation experiment, but this may not be detectable in practice.

According to Van Schaftingen, fructose 6-phosphate modulates the action of hexokinase D by favoring the formation of a complex between it and the regulatory protein, which inhibits the enzyme, whereas fructose 1-phosphate acts by preventing the formation of such a complex. As expected from this interpretation, the concentration of regulatory protein required to inhibit hexokinase D activity by 50% increases as a linear function of the enzyme concentration, because a significant amount of inhibitor then becomes trapped in the complex between hexokinase D and the regulatory protein.[38] The slope of the relationship indicates that one unit of regulator (about 5 pmol) binds to about 5 pmol hexokinase D, confirming the suggestion that there is a stoichiometric 1:1 complex between the two proteins. As fructose 1-phosphate and fructose 6-phosphate are not effectors of the purified enzyme they ought to bind either to the regulatory protein or to hexokinase D at a site that becomes available only when hexokinase D is complexed with the regulatory protein. Binding studies showing that the regulatory protein binds up to 1 mol/mol of fructose 1-phosphate

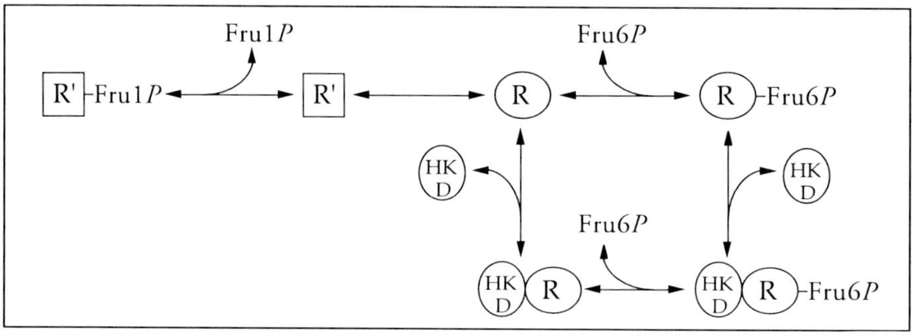

Fig. 7.9. Model for the action of the regulatory protein on hexokinase D. The regulatory protein exists in two different conformations designated R and R′. R can bind to hexokinase D and inhibits it. Effectors such as fructose 6-phosphate and sorbitol 6-phosphate, which displace the equilibrium towards R, inhibit hexokinase D in the presence of regulatory protein. Fructose 1-phosphate displaces the equilibrium towards R' and thus counteracts the inhibition. Reproduced with permission from FASEB J 1994; 8:414-419.

and sorbitol 6-phosphate, the analog of fructose 6-phosphate mentioned already, favor the first hypothesis.[46]

For the action of the regulatory protein Van Schaftingen has proposed the model shown in Figure 7.9,[22,35] in which both fructose phosphates bind to the regulatory protein and induce different conformational changes, or stabilize different conformations of the enzyme. In this model the regulatory protein exists under two different conformations, one represented by circles (R), which can bind both hexokinase D and fructose 6-phosphate, and another, represented by squares (R'), which binds fructose 1-phosphate but cannot bind hexokinase D. R' is postulated to predominate over R in the absence of ligand. Fructose 6-phosphate reinforces the inhibition by increasing the R/R' ratio, whereas fructose 1-phosphate counteracts the inhibition by decreasing it. The complex between fructose 6-phosphate and the regulatory protein inhibits hexokinase D competitively with respect to glucose. This does not mean, however, that they must bind to the same site but only that they compete for the enzyme, not being able to be bound simultaneously.

In a kinetic study using different combinations of inhibitors and studying the effect on the hexokinase D activity Van Schaftingen's group showed that *N*-acetylglucosamine inhibits synergistically with palmitoyl-CoA and with regulatory protein.[36] They concluded that the inhibitory complex binds to the enzyme at the site of the acyl-CoA rather than at the site of *N*-acetylglucosamine,[36] arguing that it would not be

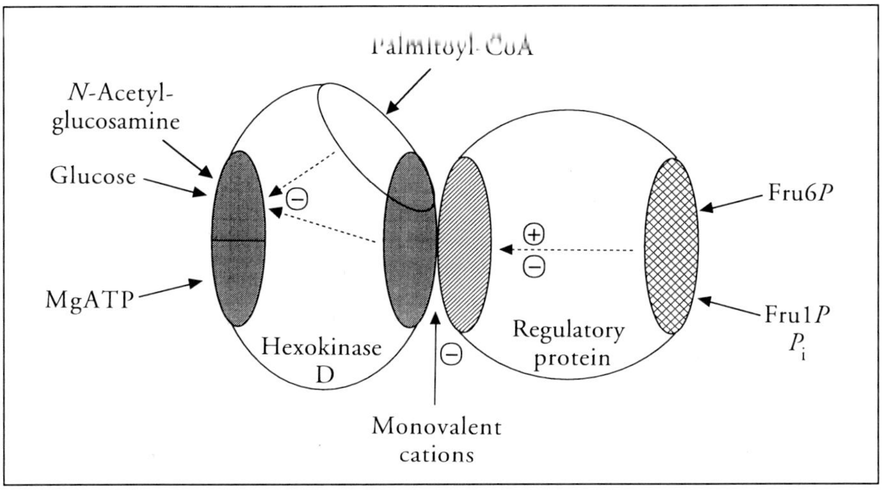

Fig. 7.10. Interactions of substrates and effectors on hexokinase D and its regulatory protein. Note that certain effectors, such as N*-acetylglucosamine, interact directly with hexokinase D, whereas others, such as fructose 6-phosphate, act indirectly by modulating the interaction of hexokinase D with the regulatory protein. Reproduced with permission from FASEB J 1994; 8:414-419.*

at the active site but at an allosteric site, as schematized in Figure 7.10. One has to be cautious, however, about localizing binding sites on the basis of kinetic data only, because the fact that long chain acyl-CoA and the regulatory protein appear to compete with each other only indicates that their binding is mutually exclusive and that they cannot form a ternary complex with the enzyme, not necessarily that they bind at the same site. Something to bear in mind is that *N*-acetylglucosamine is a very untypical inhibitor of hexokinase D, with a very high affinity for the enzyme in the presence of glucose but a very low affinity in its absence, as we discussed in Chapter 6, and this may complicate the interpretation. A difficult point to understand is that the kinetic data of Vandercammen and Van Schaftingen suggest that *N*-acetylglucosamine binds equally well to the free enzyme and to the binary complex between hexokinase D and the regulatory protein:[36] this is a little puzzling because *N*-acetylglucosamine is a glucose analog, and glucose binding appears to be affected by the regulatory protein, as the $K_{0.5}$ for glucose is greatly increased.

The fact that all vertebrate hexokinase D show inhibition by the regulatory protein and by long-chain acyl-CoAs[36] indicates that there is a large degree of conservation in the determinants for the interactions between hexokinase D and the regulatory protein on the one hand and between hexokinase D and acyl-CoA on the other. Neither of the two effects is observed with brain or heart hexokinases (hexokinases A and B respectively), nor with glucokinase of microorganisms. This is not surprising, however, as the glucokinase of microorganisms is a different enzyme with no particular correspondence with liver hexokinase D (Chapter 2).

Table 7.2. Comparative characteristics of inhibition of hexokinase D by acyl-CoAs and by the regulatory protein

	Acyl-CoA	**Regulatory protein[1]**
Inhibition at saturation	Partial[2]	Complete[3]
Inhibitor cooperativity	Cooperative	Non-cooperative
Variable [Glucose]	Competitive	Competitive
Variable [ATP]	Competitive	Mixed[4]
Effect on glucose cooperativity	No effect	Small decrease

1 Assay mixture with equimolar total Mg^{2+} and ATP, in the presence of fructose 6-phosphate.

2 Assay mixture with excess of Mg^{2+}. With equimolar total Mg^{2+} and ATP the inhibition is complete.

3 Partial in the absence of fructose 6-phosphate.

4 The behavior is non-linear, but as equimolar total Mg^{2+} and ATP concentrations were used, and as fructose 6-phosphate also complexes with Mg^{2+}, there are uncertainties in the interpretation that cannot be resolved at present.

The authors suggest that the palmitoyl-CoA site has been conserved through evolution as part of the regulatory protein binding site. This is not very clear cut, however, because of the very different chemical natures of the two inhibitors. Protein side chains do not mimic the CoA or the acyl groups very well, and one has to bear in mind the specificity in relation to the length of the acyl chain. Also there are important kinetic differences between the effects of palmitoyl-CoA and the regulatory protein (Table 7.2), even though they both show competitive inhibition with respect to glucose.

K^+ and the regulatory protein

As mentioned above, K^+ appears to stimulate glucose phosphorylation in hepatocytes, mimicking the fructose effect. Although initially it was thought that in the presence of K^+ the levels of fructose 1-phosphate could be increased, the stimulating effect of K^+ is not due to the increase in the formation of fructose 1-phosphate but to an increase of the intracellular concentration of chloride, which relieves the inhibition exerted by the regulatory protein on purified hexokinase D,[36] and to a swelling of the cells that produces a dilution of the regulatory protein.[47] These two effects are presumably antagonized to the same extent by a decrease in the intracellular concentration of inorganic phosphate, which has an effect like that of fructose 1-phosphate.

Tissue and species distribution

Like hexokinase D the regulatory protein is present in the liver and probably also in pancreatic islets, but not in brain, skeletal muscle, heart, spleen and kidney.[39] The corresponding mRNA is also absent from these tissues, which suggests that the gene is turned off. Metabolic zonation in liver, with hexokinase D predominating in the hepatocytes of the perivenous domain, as described in Chapter 4, makes it important to establish whether the regulatory protein is also distributed zonally, but this is not known at present.

It appears that the regulatory protein, like hexokinase D, is also present in β-cells of the islets of Langerhans,[48] although it is hard to be certain at this moment. Pancreatic hexokinase D is sensitive to the action of the regulatory protein of hepatic origin. Glucose phosphorylation in islets homogenates is slightly stimulated by fructose 1-phosphate, but not by fructose, which is not surprising in view of the low fructokinase activity in these cells. Glyceraldehyde, which increases the concentration of fructose 1-phosphate in the islets, stimulates glucose phosphorylation. The role of the regulatory protein in pancreatic islets remains unclear.

As we saw in Chapter 1, hexokinase D is expressed in the livers of various mammalian species, like human, rat, pig and dog, but not in cat or ruminants. The regulatory protein is found only in those species that have hexokinase D.[39] The same correspondence is observed in other vertebrate species, such as the turtle, the frog and the toad, which express both proteins, whereas the chicken expresses neither. This coexpression suggests that the two proteins form a functional unit. Interestingly, the regulatory proteins from lower vertebrates are unaffected by fructose 6-phosphate, fructose 1-phosphate or other phosphate esters known to be effectors[35] of rat regulatory protein. Despite these differences these regulatory proteins share several properties with that from rat: (i) they act as competitive inhibitors with respect to glucose; (ii) their inhibitory effect is relieved by KCl; (iii) their molecular masses are about 60-75 kDa; (iv) they are retained on anion-exchangers but are eluted at a salt concentration of 100 mM.

PHYSIOLOGICAL IMPLICATIONS OF THE REGULATORY PROTEIN

In the absence of fructose, the liver cell contains 10-50 μM fructose 6-phosphate but no detectable fructose 1-phosphate,[49] and so the regulatory protein, which gives half-maximal inhibition at about 10 nM with purified hexokinase D, can inhibit the enzyme in the liver. The apparent affinity of hexokinase D for glucose in intact hepatocytes is thus lower than that of pure hexokinase D. The fact that fructose and other compounds giving rise to fructose 1-phosphate restore the affinity to the expected level is in accordance with this interpretation.[33]

This type of regulation appears to occur also in vivo.[50] Fructose, injected via the penile vein, stimulates the rate of glucose phosphorylation up to about 2.5-fold, both in fed animals and in animals starved overnight, with maximal effects observed at a dose of 50 mg/kg. Fructose was also active when administered by intragastric infusion. It caused increases in the concentration of fructose 1-phosphate, which reached values sufficient to release the inhibition exerted by the regulatory protein. Even rather low amounts of fructose in the diet (e.g. a few grams for a human adult) are therefore likely to stimulate glucose-phosphorylation by the liver.[21] The potential importance of this is underlined by the fact that fructose, the sweetest sugar that occurs naturally in the diet, has become a major nutrient, with 50-100 g of the free monosaccharide or as sucrose consumed daily by adults in the Western world.[51]

The regulatory protein provides an explanation for the observation that fructose administration causes a moderate hypoglycemia in anesthetized rats or in perfused livers[47] and for the fact that fructose increases the conversion of radiolabeled glucose to glycogen.[23,27] What could be the physiological significance of this "collaboration" between glucose and fructose? Portal blood contains no fructose in the post-absorptive state; but as fructose together with glucose is present in most food of vegetable origin, it could be a signal that allows an increase in glucose uptake by the liver. Although flour is an exception, it is also a relative late-comer in nutritional history, in the sense that virtually all of human (and rat) evolution occurred before the invention of farming.

QUESTIONNAIRE

Receive a FREE BOOK of your choice

Please help us out—Just answer the questions below, then select the book of your choice from the list on the back and return this card.

R.G. Landes Company publishes five book series: *Medical Intelligence Unit, Molecular Biology Intelligence Unit, Neuroscience Intelligence Unit, Tissue Engineering Intelligence Unit* and *Biotechnology Intelligence Unit.* We also publish comprehensive, shorter than book-length reports on well-circumscribed topics in molecular biology and medicine. The authors of our books and reports are acknowledged leaders in their fields and the topics are unique. Almost without exception, there are no other comprehensive publications on these topics.

Our goal is to publish material in important and rapidly changing areas of bioscience for sophisticated scientists. To achieve this goal, we have accelerated our publishing program to conform to the fast pace in which information grows in bioscience. Most of our books and reports are published within 90 to 120 days of receipt of the manuscript.

Please circle your response to the questions below.

1. We would like to sell our *books* to scientists and students at a deep discount. But we can only do this as part of a prepaid subscription program. The retail price range for our books is $59-$99. Would you pay $196 to select four *books* per year from any of our Intelligence Units–$49 per book–as part of a prepaid program?

   Yes **No**

2. We would like to sell our *reports* to scientists and students at a deep discount. But we can only do this as part of a prepaid subscription program. The retail price range for our reports is $39-$59. Would you pay $145 to select five *reports* per year–$29 per report–as part of a prepaid program?

   Yes **No**

3. Would you pay $39–the retail price range of our books is $59-$99–to receive any single book in our Intelligence Units if it is spiral bound, but in every other way identical to the more expensive hardcover version?

   Yes **No**

To receive your free book, please fill out the shipping information below, select your free book choice from the list on the back of this survey and mail this card to:

R.G. Landes Company, 909 S. Pine Street, Georgetown, Texas 78626 U.S.A.

Your Name ______________________________

Address ______________________________

City ______________________ State/Province: ______________________

Country: ______________________ Postal Code: ______________________

My computer type is Macintosh ______ ; IBM-compatible ______ ; Other ______

Do you own ____ or plan to purchase ___ a CD-ROM drive?

THE REGULATORY PROTEIN AND THE GLUCOSE PARADOX

It has been generally accepted for many years that muscle and liver glycogen are both synthesized from glucose via a direct pathway (glucose → glucose 6-phosphate → UDPglucose → glycogen) without intermediate degradation of the carbon chain of glucose. However, a few earlier studies and some more recent ones have provided evidence that liver glycogen is synthesized primarily via the indirect pathway (glucose → glucose 6-phosphate → pyruvate → glucose 6-phosphate → UDPglucose → glycogen), i.e. with glycolytic degradation of the glucose chain to pyruvate followed by gluconeogenic resynthesis. These findings were termed the "glucose paradox."[52] In primary cultures of hepatocytes the pathway of glycogen synthesis appeared to depend on the prior metabolic state of the animal,[53] as hepatocytes isolated from fed rats synthesized glycogen via the direct pathway whereas those from fasted rats used the indirect pathway.

In many of the fasting and refeeding experiments the animals were fasted and then given glucose intragastrically, or the livers were perfused with glucose,[52,54-56] and consequently no fructose was present. In these conditions, as mentioned above, the amount of fructose 6-phosphate in the hepatocytes appears to be sufficient to inhibit hexokinase D by interaction with the regulatory protein, and the $K_{0.5}$ for glucose would be substantially increased. It will be very important therefore to repeat studies of this kind in the presence of fructose, which greatly increases the rate of glycogen synthesis, as mentioned earlier in this chapter, and also activates hexokinase D by liberating it from inhibition by the regulatory protein. This regulation could not only inhibit the hexokinase D activity but could also disturb putative interactions with other enzymes, such as phosphoglucomutase for example.

COUNTERPOINT BETWEEN PALMITOYL-COA AND REGULATORY PROTEIN AS EFFECTORS IN VIVO OF HEXOKINASE D

As mentioned above, the physiological significance of palmitoyl-CoA inhibition of hexokinase D is still inconclusive, whereas the role of Van Schaftingen's regulatory protein appears clearer, as it explains some phenomena in hepatocytes. That does not mean, however, that there is no role for palmitoyl-CoA. Although the results of Van Schaftingen and Vandercammen with oleate appear to cast serious doubts on the role of long-chain acyl-CoA, one must consider the possibility that the amount of these compounds produced by isolated hepatocytes may be limited. The cellular studies are difficult to do definitively, and lipolysis may be insufficient. An important point to bear in mind is that in conditions of intense lipolysis such as diabetes or starvation fatty acids arrive in the liver in substantial amounts and although one may see no effect in isolated hepatocytes the effect may exist in the living animal, when the liver is perfused with blood. In fact, the earlier animal studies of Seitz,[57,58] and even Hue's cellular studies,[59] suggested that increasing levels of acyl-CoA could inhibit hexokinase D or perhaps glycolysis. The allosteric effect of palmitoyl-CoA on hexokinase D would be quite straightforward. Both modes of regulation could be used in different metabolic conditions, for example lipolysis for acyl-CoA and fructose availability for the regulatory protein. In relation to this it is important to remember that insulin is sufficient for the induction of hexokinase D in the isolated hepatocyte, but not in the whole animal.

In relation to the mechanisms of action of acyl-CoA and the regulatory protein, it is not clear if they bind at the same site or not; what is clear is that they cannot bind at the same time, i.e. they cannot be both together bound to the same hexokinase D molecule. In view of the very different chemical natures of the two inhibitors one is more inclined to think of different sites, or of sites that overlap only partially, as Van Schaftingen does. As there are many different kinetic forms of hexokinase D it would be easy to imagine a subset that binds the regulatory protein but not acyl-CoA and vice versa. The modest effect of the regulatory protein[36] on the cooperativity still appears to be greater than that of palmitoyl-CoA,[14] and much less than that of *N*-acetylglucosamine.[17] This may suggest that the regulatory protein binds to a subset that shifts the distribution of forms at the steady state slightly, much less than mannose or *N*-acetylglucosamine, whereas palmitoyl-CoA does not shift the distribution because it binds equally well to all kinetically significant forms.

Now that both hexokinase D and the regulatory protein have been cloned the next step is to engineer a hexokinase D that is insensitive to palmitoyl-CoA or to the regulatory protein by mutating the allosteric binding site and thus proving this hypothesis. The use of transgenic mice carrying insensitive forms of hexokinase D would allow the roles of palmitoyl-CoA and the regulatory protein to be dissected.

References

1. Stadtman ER. Mechanisms of enzyme regulation in metabolism. In: Boyer PD, ed. The Enzymes, 3rd ed. New York: Academic Press 1970; 1:397-459.
2. Dawson CM, Hales CN. The inhibition of rat liver glucokinase by palmitoyl-CoA. Biochim Biophys Acta 1969;176:657-659.
3. Clark-Turri L, Peñaranda J, Rabajille E et al. Immunochemical titration of liver glucokinase from normal, fasted and diabetic rats. FEBS Lett 1974; 41:342-344.
4. Weissmann C. A phage in New York. In: Kornberg A, Horecker BL, Cornudella L et al., eds. Reflections on Biochemistry in Honour of Severo Ochoa. Oxford: Pergamon, 1976:283-292.
5. Niemeyer H, Cerpa C, Rabajille E. Inhibition of hexokinase activity by a fructose 2,6-bisphosphate-dependent cytosolic protein from liver. Arch

Biochem Biophys 1987; 257:17-26.

6. Niemeyer H, Rabajille E. Phosphofructokinase is responsible for fructose 2,6-bisphosphate inhibition of hexokinase in tissue extracts. Arch Biochem Biophys 1988; 258:91-93.
7. Van Schaftingen E. A protein from rat liver confers to glucokinase the property of being antagonistically regulated by fructose 6-phosphate and fructose 1-phosphate. Eur J Biochem 1989; 179:179-184.
8. Ekman P, Nilsson E. Phosphorylation of glucokinase from rat liver in vitro by protein kinase A with a concomitant decrease of its activity. Arch Biochem Biophys 1988; 261:275-282.
9. Kemp BE, Pearson RB. Protein kinase recognition sequence motifs. Trends Biochem Sci 1990; 15:342-346.
10. Iynedjian PB. Mammalian glucokinase and its gene. Biochem J 1993; 293:1-13.
11. Tubbs PK, Garland PB. Variations in tissue contents of coenzyme A thio esters and possible metabolic implications. Biochem J 1964; 93:550-557.
12. Srere P. Palmitoyl-coenzyme A inhibition of the citrate-condensing enzyme. Biochim Biophys Acta 1965; 106:445-455.
13. Taketa K, Pogell BM. The effect of palmitoyl coenzyme A on glucose 6-phosphate dehydrogenase and other enzymes. J Biol Chem 1966; 241:720-726.
14. Tippett PS, Neet KE. Specific inhibition of glucokinase by longchain acyl coenzymes A below the critical micelle concentration. J Biol Chem 1982; 257:12839-12845.
15. Tippett PS, Neet KE. An allosteric model for the inhibition of glucokinase by long chain acyl coenzyme A. J Biol Chem 1982; 257:12846-12852.
16. Cárdenas ML, Rabajille E, Niemeyer, H. Fructose: a good substrate for rat-liver "glucokinase" (hexokinase D). Biochem J 1984; 222:363-370.
17. Cárdenas ML, Rabajille E, Niemeyer, H. Suppression of kinetic cooperativity of hexokinase D ("glucokinase") by competitive inhibitors. Eur J Biochem 1984; 145:163-171.
18. Neet KE, Tippett PS, Keenan RP. Regulatory properties of glucokinase. In: Chock PB, Huang CY, Tsou CL et al., eds. Enzyme Dynamics and Regulation eds. New York: Springer-Verlag, 1988:28-38.
19. Lin SX, Vogel WK, Neet KE. Stoichiometry of slow binding of palmitoyl-CoA to liver glucokinase. Int J Pept Prot Res 1989; 34:333-339.
20. Woldegiorgis G, Spennetta T, Corkey BE et al. Extraction of tissue long-chain acyl-CoA esters and measurement by reverse-phase high-performance liquid chromatography. Anal Biochem 1985; 150:8-12.
21. Van Schaftingen E, Vandercammen A, Detheux M et al. The regulatory protein of liver glucokinase. Advan Enzyme Regul 1992; 32:133-148.
22. Van Schaftingen E, Detheux M, Veiga da-Cunha M. Short-term control of glucokinase activity: role of a regulatory protein. FASEB J 1994; 8:414-419.
23. Seglen PO. Autoregulation of glycolysis, respiration, gluconeogenesis and glycogen synthesis in isolated parenchymal rat liver cells under aerobic and anaerobic conditions. Biochim Biophys Acta 1974; 338:317-336.
24. Figueroa E, Niemeyer H, González C. Influence of DL-glyceraldehyde and

of L-sorbose-1-phosphate on glycogen synthesis from glucose by rat liver slices. Acta Physiol Latinoam 1956; 6:112-116.
25. Parmiak MA, Kalant N. Enhancement of glycogen concentrations in primary cultures of rat hepatocytes exposed to glucose and fructose. Biochem J 1988; 251:795-802.
26. Youn JH, Kaslow HR, Bergman RN. Fructose effect to suppress hepatic glycogen degradation. J Biol Chem 1987; 262:11470-11477.
27. Clark DG, Filsell OH, Topping DL. Effects of fructose concentration on carbohydrate metabolism, heat production and substrate cycling in isolated rat hepatocytes: Biochem J 1979; 184:501-507.
28. Topping DL, Mayes PA. Comparative effects of fructose, glucose and lactate in the splanchnic blood vessels of rats absorbing fructose. Nutr Metab 1971; 13:331-338.
29. Katz J, Rognstad R. Futile cycles in the metabolism of glucose. Curr Top Cell Regn 1976; 10:237-289.
30. Cárdenas ML, Rabajille E, Niemeyer H. Kinetic cooperativity of glucokinase with glucose. Arch Biol Med Exp 1979; 12:571-580.
31. Bontemps F, Hue L, Hers H-G. Phosphorylation of glucose in isolated rat hepatocytes. Sigmoidal kinetics explained by the activity of glucokinase alone. Biochem J 1978; 174:603-611.
32. Hue L, Bontemps F, Hers H-G. The effect of glucose and of potassium ions on the interconversion of the two forms of glycogen phosphorylase and glycogen synthetase in isolated rat liver preparations. Biochem J 1975; 152:105-114.
33. Van Schaftingen E, Vandercammen A. Stimulation of glucose phosphorylation by fructose in isolated rat hepatocytes. Eur J Biochem 1989; 179:173-177.
34. Niemeyer H, Cárdenas ML, Rabajille E et al. Sigmoidal kinetics of glucokinase. Enzyme 1975; 20:321-333.
35. Detheux M, Vandercammen A, Van Schaftingen, E. Effectors of the regulatory protein acting on liver glucokinase: a kinetic investigation. Eur J Biochem 1991; 200:553-561.
36. Vandercammen A, Van Schaftingen E. Competitive inhibition of liver glucokinase by its regulatory protein. Eur J Biochem 1991; 200:545-551.
37. Van Schaftingen E, Hue L, Hers H-G. Extracellular metabolites in suspensions of isolated hepatocytes. Biochem J 1987; 248:517-521.
38. Vandercammen A, Van Schaftingen E. The mechanism by which rat liver glucokinase is inhibited by the regulatory protein. Eur J Biochem 1990; 191:483-489.
39. Vandercammen A, Van Schaftingen E. Species and tissue distribution of the regulatory protein of glucokinase. Biochem J 1993; 294:551-556.
40. Detheux M, Vandekerckhove J, Van Schaftingen E. Cloning and sequencing of rat liver cDNAs encoding the regulatory protein of glucokinase. FEBS Lett 1993; 321: 111-115; corrected in FEBS Lett 1994; 339:312.
41. White TK, Wilson JE. Isolation and characterization of the discrete *N*- and *C*-terminal halves of rat brain hexokinase: retention of full catalytic

activity in the isolated *C*-terminal half. Arch Biochem Biophys 1989; 174:375-393.

42. Magnani M, Bianchi M, Casabianca A et al. A recombinant human "mini"-hexokinase is catalytically active and regulated by hexose 6-phosphates. Biochem J 1992; 285:193-199.
43. Arora KK, Filburn CR, Pedersen PL. Structure/function relationships in hexokinase. Site-directed mutational analyses and characterization of overexpressed fragments implicate different functions for the *N*- and *C*-terminal halves of the enzyme. J Biol Chem 1993; 268:18259-18266.
44. Crane RK, Sols A. The non-competitive inhibition of brain hexokinase by glucose-6-phosphate and related compounds. J Biol Chem 1954; 210:597-606.
45. Cárdenas ML, Rabajille E, Niemeyer H. Maintenance of the monomeric structure of glucokinase under reacting conditions. Arch Biochem Biophys 1978; 190:142-148.
46. Vandercammen A, Detheux M, Van Schaftingen E. Binding of sorbitol-6-phosphate and of fructose 1-phosphate to the regulatory protein of liver glucokinase. Biochem J 1992; 286:253-256.
47. Van Schaftingen E, Vandercammen A. Mechanism of the stimulatory effect of a potassium-rich medium on the phosphorylation of glucose in isolated rat hepatocytes. Eur J Biochem 1992; 204:363-369.
48. Malaisse WJ, Malaisse-Lagae F, Davies DR et al. Regulation of glucokinase by a fructose-1-phosphate-sensitive protein in pancreatic islets. Eur J Biochem 1990; 190:539-545.
49. Davies DR, Detheux M, Van Schaftingen E. Fructose-1-phosphate and the regulation of glucokinase activity in isolated hepatocytes. Eur J Biochem 1990; 192:283-289.
50. Van Schaftingen E, Davies DR. Fructose administration stimulates glucose phosphorylation in the livers of anesthetized rats. FASEB J. 1991; 5:326-330.
51. Cox T. Aldolase B and fructose intolerance. FASEB J. 1994; 8:62-70.
52. Katz J, Kuwajima M, Foster DW et al. The glucose paradox: new perspectives on hepatic carbohydrate metabolism. Trends Biochem Sci 1986; 11:136-140.
53. Spence JT, Koudelka AP. Pathway of glycogen synthesis from glucose in hepatocytes maintained in primary culture. J Biol Chem 1984; 260:1521-1526.
54. Hems DA, Whitton PD, Taylor EA. Glycogen synthesis in the perfused liver of the starved rat. Biochem J 1972; 129:529-533.
55. Newgard CB, Hirsch LJ, Foster DW et al. Studies on the mechanism by which exogenous glucose is converted into liver glycogen in the rat. A direct or indirect pathway? J Biol Chem 1983; 258:8046-8050.
56. Newgard CB, Moore SU, Foster DW et al. Efficient hepatic glycogen synthesis in refeeding rats requires continued carbon flow through the gluconeogenic pathway. J Biol Chem 1984; 259:6958-6963.
57. Seitz HJ, Müller MJ, Krone W et al. Coordinate control of intermediary metabolism in rat liver by the insulin/glucagon ratio during starvation and after refeeding. Regulatory significance of long-chain acyl-CoA and

cyclic AMP. Arch Biochem Biophys 1977; 183:647-663.
58. Seitz HJ, Müller MJ, Krone W et al. Rapid conversion by insulin of hepatic intermediary metabolism from glucose production to glucose utilization in the liver of alloxan-diabetic rats. Diabetes 1977; 26:1159-1174.
59. Hue L, Maisin L, Rider MH. Palmitate inhibits liver glycolysis. Involvement of fructose 2,6-bisphosphate in the glucose/fatty acid cycle. Biochem J 1988; 251:541-545.

CHAPTER 8

CONTROL OF GENE EXPRESSION

As discussed in Chapter 1, studies in the whole animal showed that the hexokinase D activity in rat liver was dependent on the simultaneous presence of insulin and glucose. On the other hand glucagon, epinephrine and cyclic AMP appeared to inhibit induction of hexokinase D. In addition other hormones such as glucocorticoids appear to modulate the induction rate. All the data suggested variation of gene expression in the different conditions, with regulation probably exerted at the transcription level. However, the mechanism of regulation of the expression of the hexokinase D gene was far from clear, and there were several important questions unsolved. Which was the real inducer, glucose or insulin, or both? Why was cyclic AMP such a powerful inhibitor of the induction of hexokinase D, capable of predominating over the combined action of insulin and glucose? What mechanisms are involved?

Another question arose from the proof that hexokinase D was present in β-cells and the characterization of this form of the enzyme. Although hepatic and pancreatic hexokinase D have similar kinetic behavior (Chapters 3 and 6), they respond differently to the nutritional state of the animal, the pancreatic enzyme being largely constitutive. What is the reason for this difference in regulation? The discovery of different promoter regions in a single hexokinase D gene, with tissue-specific expression (Chapter 5) was crucial for understanding the differences in regulation between the hepatic and islet isoforms.

Understanding the mechanism responsible for regulating expression of the hexokinase D gene was not easy. Nevertheless, some of the questions have now been answered, as we shall see in this chapter. Several valuable reviews have been published recently by groups that have contributed greatly to the elucidation of hexokinase D gene expression.[1-4]

FIRST STUDIES WITH ISOLATED HEPATOCYTES

The whole animal, as used in the early studies, is obviously a very complex system for establishing the molecular basis of the participation of glucose and each of the different hormones involved in maintaining the level of hexokinase D activity. As indicated before, glucose injection in vivo is followed by an increase in plasma insulin and a decrease in plasma glucagon concentrations. Interpretation of studies in the whole animal is therefore complicated by complex feedback regulatory mechanisms involving multiple effectors. Primary cultures of hepatocytes do not present this complexity and constitute an excellent system for studying the induction of hexokinase D. Consequently, it is not surprising that as early as the 1970s the most suitable system to study the role of individual hormones appeared to be the isolated hepatocyte.

Nevertheless, when the studies in isolated hepatocytes were finally done, at the end of the 1970s and early 1980s, an unexpected feature emerged. The previous studies on the whole rat done by different groups had given very coherent results on the whole, with differences that were more quantitative than qualitative. By contrast, the first studies with hepatocytes produced some serious disagreements between the results of the different groups, even though in general they were consistent with some of the studies on the whole animal; for example, in many of these studies the induction of hexokinase D was prevented by glucagon and cyclic AMP.[5-8] The discrepancies are such that if one examines the corresponding literature of that period, an important proportion of the information could be considered very chaotic, especially in relation to the role of glucose and insulin. The disagreements probably had their origin in differences in how the hepatocytes were prepared and cultured, as the conditions were not well standardized. Whatever the reason, the molecular mechanism responsible for the induction of hepatic hexokinase D was far from being understood at the beginning of the 1980s.

Something new that emerged in this period from the studies in cultured hepatocytes was the participation of cyclic GMP (cGMP). Thus, Spence and colleagues found a biphasic increase in activity after insulin addition to primary cultures of hepatocytes (a curve qualitatively similar to Figure 1.3,[9] the second phase being dependent on the presence of carbohydrates or cGMP. 8-Bromo-cGMP stimulated synthesis of hexokinase D to an extent similar to that observed with insulin, and the effects were additive.[9]

INDUCTION OF HEXOKINASE D mRNA

The development in the 1980s of new and more sensitive techniques, such as the Western and Northern blots, allowed the detection of enzymes and their mRNA with greater sensitivity. Cloning and sequencing of the hexokinase D gene allowed the assembly of a new set of information, especially in relation to the mRNA. It was first reported that the variations in hexokinase D activity were related to the concentrations of the specific mRNA,[10] and Iynedjian's group later confirmed and extended these results in studies in the whole animal. They found that hexokinase D mRNA is undetectable by Northern blot in total liver RNA from rats fasted for 24-72 h.[11] As expected, oral administration of glucose to such animals causes a rapid, massive and transient accumulation of the hepatic 2.5 kb mRNA for hexokinase D. Induction of the mRNA culminates 6-10 h after refeeding in a level as high as that seen in the livers of normal animals fed *ad libitum*.[11] In contrast, the typical 2.8 kb mRNA for hexokinase D is maintained at a constant level in the islets of Langherhans even during prolonged starvation. Moreover, there is no increase in islet mRNA after an oral glucose load,[12] in agreement with the rather modest variation of hexokinase D activity observed in pancreas.[13] As expected, Western blotting failed to show any significant change in the amount of hexokinase D protein in pancreatic islets during the fasting-refeeding transition, whereas the liver enzyme increased 3-fold within 18 h of glucose refeeding,[12] suggesting that tissue-specific transcription mechanisms result in inducible expression of the gene in liver but not in islets during the fasting-refeeding transition.

All these new results reinforce the idea that transcription of the hexokinase D gene in liver is turned on by a nutritional signal, but that in contrast the islet β-cell promoter is unresponsive. Transcription of the gene for hexokinase D in islet β-cells does not appear to be regulated in the whole animal by hormones or other effectors, as the variations in enzyme activity and immunoreacting protein are minor. In addition, insulin has no effect on expression of hexokinase D in cultured rat islets, but glucose does have an effect,[14] as hexokinase D activity varied by about five-fold in islets cultured for 7 days at different glucose concentrations, and the magnitude of the effect was concentration-dependent. However, the mechanism involved in the regulation of hexokinase D by glucose may actually involve stabilization of the enzyme protein or translational control, possibly related to the unique 5' untranslated region of the mRNA in the β-cell, as the level of hexokinase D mRNA was unchanged under these conditions.[2,14] Down-regulation of islet hexokinase D mRNA as well as proinsulin mRNA after prolonged exercise training has been reported,[15] which may suggest coordinate regulation, especially in view of the fact that no change in islet glucose transporter (GLUT2) mRNA was detected. However, the decrease in enzyme mRNA was accompanied by a similar decrease in total RNA content of the islet, which made interpretation of these results more difficult. The discovery of the existence of different promoter regions in a single hexokinase D gene (Chapter 5), tissue-specific in their expression, was crucial for understanding the differences in regulation between the liver and islet isoforms. The enzyme in the islets of Langerhans is basically expressed constitutively, regardless of the nutritional status of the animal, whereas in the liver it behaves as a typical adaptive enzyme.

Thus it appears that the β-cell promoter is not acutely regulated by hormones or other signals, though this remains to be established definitively. As the structure of the β-cell specific promoter has been dissected into discrete binding sites for regulatory *trans*-acting factors,[1] the characterization of these proteins and the cloning of their genes will provide important clues in this respect.

WHAT TURNS ON THE SWITCH: GLUCOSE OR INSULIN?

As we have already discussed in Chapter 1, the presence of glucose is not by itself sufficient for inducing hexokinase D, and the response of the liver enzyme to glucose refeeding is abolished in animals simultaneously treated with anti-insulin serum.[16] On the other hand, insulin alone does not induce hepatic hexokinase D activity in diabetic rats fed a carbohydrate-free diet.[16] All of this suggested that the simultaneous presence of insulin and glucose was required.

A role for insulin as positive effector of hepatic hexokinase D expression in the whole animal is further illustrated in diabetes mellitus, as the mRNA is absent from the livers of streptozotocin-diabetic rats. Insulin treatment is followed by a prompt build-up of the mRNA, with a marked overshoot above the normal level for 10 h after the first injection of insulin.[17] In these experiments hexokinase D was also absent from the livers of diabetic rats, in contrast to what we observed in rats with 4 days of diabetes.[18] Thus in rats with streptozotocin-induced diabetes, with a final glycemia of at least 350 mg of glucose per 100 ml of blood, levels of hexokinase D were about 15% of control levels.

INDUCTION OF HEXOKINASE D mRNA BY INSULIN IN THE WHOLE ANIMAL

Run-on transcription assays with isolated nuclei provided direct evidence for a transcriptional activation of the hexokinase D gene after administration of insulin to diabetic animals.[17] However, as mentioned above, complex feedback mechanisms involving multiple effectors make it difficult to arrive at conclusions in studies on the whole animal. As a matter of fact glucose refeeding induces a transient accumulation of the mRNA, but glucose ingestion is followed by an increase in the concentration of insulin in the plasma and a drop in the level of glucagon, and both changes may conceivably contribute to hexokinase D induction in vivo. Equally, diabetes mellitus not only implies a deficiency in insulin, but also a hyperglucagonemia, and as both abnormalities are corrected by insulin treatment it is difficult to distinguish between direct effects of insulin itself and those of a decrease in the plasma glucagon level.

Insulin is the Inducer in Cultured Hepatocytes, not Glucose

So the question is whether insulin induces hexokinase D mRNA in cultured liver cells and if so whether this effect is antagonized by glucagon. Hepatocytes isolated from fasted animals and maintained in basal medium with glucose are devoid of hexokinase D mRNA, but addition of insulin to the medium at physiological concentrations elicits an immediate time-dependent increase in specific mRNA, as seen in Figure 8.1. Insulin acts at the transcriptional level, as demonstrated by run-on assays with hepatocyte nuclei,[19] and the magnitude of the effect is independent of the extracellular glucose concentration. Thus the induced level of hexokinase D mRNA 8 h after addition of insulin to the cultures was practically identical over the entire range of glucose concentrations tested, i.e. from 0.6 mM to 40 mM (inset to Fig. 8.1),[19] and similar results were obtained independently by another group.[20] Furthermore, hexokinase D mRNA was undetectable in all cells unless insulin was added, regardless of the glucose concentration in the medium. The presence of glucose appears not to alter the time course of the induction either.[19] It follows that in isolated hepatocytes the ambient glucose concentration is not a crucial factor for insulin-dependent induction of hexokinase D mRNA, and that insulin alone,

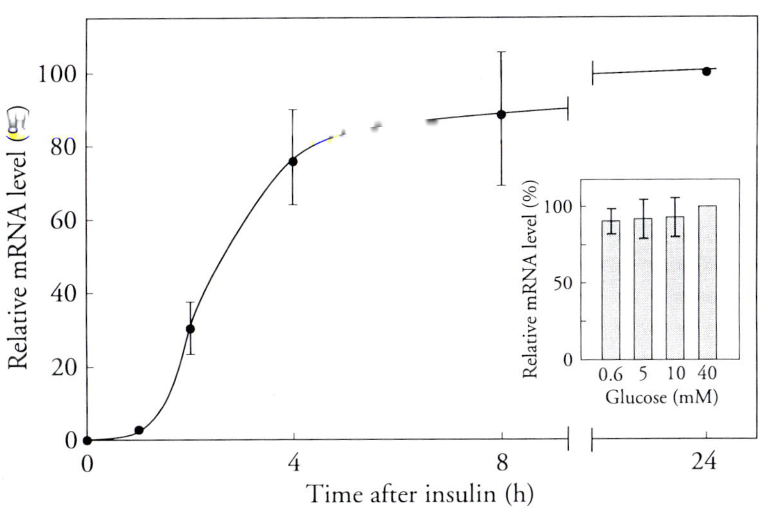

Fig. 8.1. Effect of insulin on the development of hexokinase D mRNA. Insulin was added to cultured hepatocytes isolated from fasted animals that had been maintained in basal medium containing glucose but devoid of insulin, and the time course is shown for the appearance of mRNA. The level of mRNA after 24 h incubation with insulin is taken as 100%. The inset shows the level of mRNA attained after 8 h in the presence of various concentrations of glucose. Redrawn from Figs. 1 and 5 of J Biol Chem 1989; 264: 21824-21829.

added at physiological concentrations to cultured hepatocytes, causes de novo induction of hexokinase D mRNA.[19,20] In this respect, the hexokinase D gene stands apart from a group of other genes, such as the L-type pyruvate kinase and S-14 genes, which require both insulin and high glucose concentration for transcriptional activation.[20-22] This allows us to rule out the possibility of a common operon for hexokinase D and pyruvate kinase that was mentioned in Chapter 1.

Glucagon added to hepatocytes together with a supramaximal concentration of insulin prevents induction of hexokinase D mRNA in a dose-dependent manner. When both hormones are present at maximal concentrations, the negative effect of glucagon prevails and the gene is repressed (see below). The transcriptional activity of the gene as measured by run-on assay was completely turned off within 30 min after addition of an analog of cAMP.[19] In these experiments, under deinduction conditions, a half-life for hexokinase D mRNA of about 40 min was estimated.

Is the effect of insulin a primary effect?

Another question about the activation of the hexokinase D gene by insulin is whether it is a primary effect, or whether it might be mediated by an insulin-inducible "early gene," whose newly synthesized protein product would in turn stimulate transcription of the gene. When this question was addressed, it was found that insulin induction of hexokinase D mRNA was largely or totally suppressed in hepatocytes cultured in the presence of cycloheximide, anisomycin or pactamycin.[20,23] However, a slight stimulation of specific gene transcription was still detectable in these cells by run-on assay. The interpretation of the data was further complicated by a non-specific negative effect of the inhibitors on general transcription.[23] Moreover, inhibitors of protein synthesis activate protein kinases in cultured cells and interfere with signal transduction pathways.[24,25] Taken together, the available data suggest that the effect of insulin on the hexokinase D gene is at least in part independent of concomitant protein synthesis. Specific activation of the hexokinase D gene as early as 30 min after hormone addition[1] supports the idea that the gene is a primary target for insulin action. However, the situation is different in hepatocytes from suckling rats, as we shall see later in the chapter.

REPRESSION BY GLUCAGON

Another central aspect of the regulation of hepatic hexokinase D is the acute repressor effect of cyclic AMP on the gene, as we have just seen. Interestingly, the negative effect of glucagon is not mimicked by amylin, a polypeptide produced by the pancreatic β-cell and able to counteract the effects of insulin in some target cells.[26] As cAMP exerts dominant negative control over hexokinase D gene transcription, it is legitimate to ask whether insulin might induce the gene by relieving it from basal level cAMP repression. Insulin can antagonize cAMP in liver and adipose tissue by activation of a low-K_m cyclic nucleotide phosphodiesterase termed type III phosphodiesterase.[27] At a more distant level in the signal transduction pathway, insulin can oppose cyclic AMP dependent protein phosphorylation by the stimulation of serine/threonine protein phosphatases.[28,29] Iynedjian's group has shown that insulin activation of the hexokinase gene in cultured hepatocytes is prevented by several inhibitors of the cyclic nucleotide phosphodiesterases, in particular by a preferential inhibitor of type III phosphodiesterase, and the inductive effect of insulin was suppressed in presence of low concentrations of okadaic acid, a specific inhibitor of protein phosphatases PP1 and PP2A.[23] These observations underline the importance of the interaction between insulin and the cAMP signalling system for the control of the hepatic hexokinase D gene. Iynedjian proposes a simplified scheme of the hormonal interactions in the control of hexokinase D gene transcription (Fig. 8.2), which ought to be regarded as a minimal model, as the mechanism involved in the effect of insulin remains obscure, despite identification of an abundance of potential regulatory proteins.[1,30] This field is very active at present,[30,31] and one may expect to see major advances in understanding it in the next few years. Glucagon acting via a cAMP-dependent protein kinase is shown to phosphorylate and thereby inactivate a *trans*-acting factor essential for hexokinase D gene transcription. Conversely, the putative factor is dephosphorylated and converted to the active form by an insulin-activated phosphatase. At the gene level, this implies the presence of a single hormone-response element as the target for both the positive effect of insulin and the negative effect of cAMP. It is also possible that insulin and cAMP regulate the gene via separate regulatory proteins and hormone-response elements. To distinguish between these possibilities it will be necessary to identify one or several DNA regulatory elements. However, the *cis*-regulatory elements that mediate the effects of insulin and glucagon on hexokinase D gene transcription in the liver have not been defined. Furthermore, the enhancer elements that target expression of the downstream promoter in liver have not yet been identified.[2]

REGULATION OF INITIAL EXPRESSION OF THE HEXOKINASE D GENE DURING DEVELOPMENT

NUTRITIONAL AND HORMONAL CHANGES DURING DEVELOPMENT

Birth and the suckling-weaning transition are accompanied in the rat by marked changes of nutrition, which bring about considerable hormonal modifications (see the review of Girard and colleagues,[4] and references therein). During pregnancy the fetus receives, through the placenta, a continuous supply of carbohydrates and aminoacids, but a very limited amount of free fatty acids; fetal nutrition is thus a high-

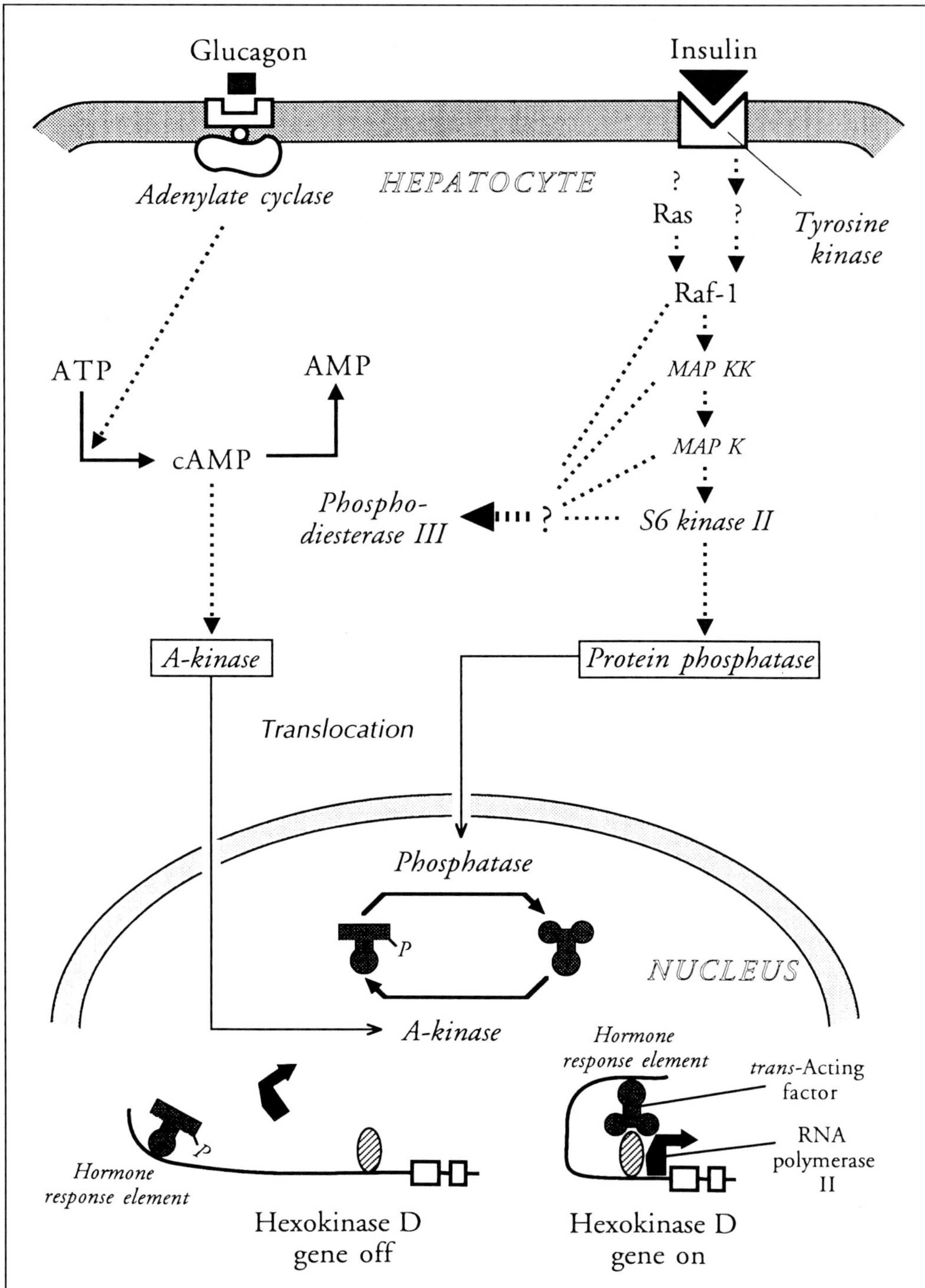

Fig. 8.2. Simplified model of hormonal interactions in the control of hexokinase D gene transcription. Adapted from Fig. 2 of Biochem J 1993; 293:1-13.

carbohydrate low-fat diet. After birth, the diet changes drastically, as during the suckling period the neonate is fed with milk, a high-fat low-carbohydrate diet. At weaning, milk is progressively replaced by rat chow, a high-carbohydrate low-fat diet, and both birth and weaning are accompanied by important changes in the hormone level. Thus, a fall in the concentration of insulin and a dramatic increase in that of glucagon occur in the plasma after birth, conditions that persist during suckling. But weaning is accompanied by a progressive increase in the concentration of insulin in plasma and a progressive decrease in the concentration of glucagon. There are also increases in the plasma corticosterone, thyroxine and triiodothyronine concentrations, beginning at about 8-10 days after birth.

In view of what we saw in Chapter 1 it is not surprising, then, that the nutritional and hormonal variations occurring during the perinatal period and at weaning would be followed by marked changes in the activities of enzymes such as hexokinase D. In fact hexokinase D offers a very good model for studying the molecular mechanisms involved in the regulation (transcriptional or post-transcriptional) of gene expression at birth and weaning, and the specific roles of nutrition and pancreatic hormones.

Changes of hexokinase D activity during development

It has long been well established that hexokinase D is absent from the livers of fetal and neonatal rats, though the other hexokinases are

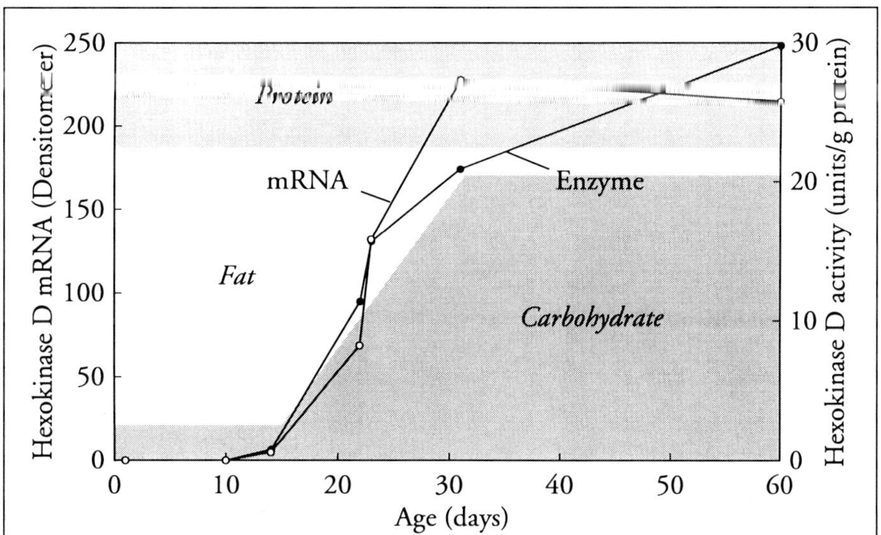

Fig. 8.3. Liver hexokinase D mRNA and activity during development of neonatal rats. The development of mRNA and enzyme activity is essentially simultaneous between the ages of 15 and 30 days. As indicated by the background shading, this development corresponds to the change of the animals from a high-fat to a high-carbohydrate diet. Plotted from data tabulated by Iynedjian and coworkers.[11]

present.[32-35] Hexokinase D activity develops spontaneously in the rat during the third week of extrauterine life and rises progressively to adult levels by the end of weaning, at about 30 days after birth.[32,34] Iynedjian's group showed that the rise of hexokinase D activity occurs concomitantly with the accumulation of specific mRNA (Fig. 8.3),[11] in agreement with Spence's results,[10] reinforcing the idea that the appearance of enzyme activity reflects the turning on of specific gene transcription.

As inactive isoforms of hexokinase D are produced by alternative splicing of hexokinase D mRNA within the liver itself (Chapter 5),[36] a question to ask at this point is whether at the beginning of the neonatal period the gene is being transcribed, but because of differential splicing is producing not the active isoform B1 but the inactive form B2 or another that is equally inactive. The fact that no hexokinase D mRNA can be detected in this period, which is now well confirmed,[4] indicates that a mechanism involving differential splicing during development can be excluded.

Hexokinase D contrasts with the other three hepatic isoenzymes, whose activities increase progressively during the first week of extrauterine life to levels about 4-fold higher than in adult rat liver and then decline, reaching their characteristic adult levels at the time of weaning.[35]

To understand the regulation of hexokinase D gene expression during development, especially in relation to the action of glucose and insulin, experimental attempts have been made to advance or to delay the age of appearance of enzyme activity. A complication in this type of experiment, however, is that suckling rats begin to nibble the mother's food from about 15 days of age, which means that in practice the transition in diet at weaning is gradual, as illustrated in Figure 8.3, and consequently there is some uncertainty in establishing with precision the time course of the increase of hexokinase D activity or mRNA after weaning to a high-carbohydrate diet.

Premature or delayed induction of hexokinase D activity in neonatal rats

Hexokinase D activity can be induced prematurely by artificially rearing 2-day-old rats with a high-carbohydrate formula via an intragastric cannula.[37] In this case the levels of hepatic hexokinase D activity reached 30% of adult activity 4 days after the start of high-carbohydrate feeding, and 70% after 10 days. Premature induction of hexokinase D activity to 30% of adult levels can also be elicited by oral administration of glucose to 10-day old rats. Interestingly, in this case the induction is prevented by simultaneous injection of mannoheptulose, which inhibits insulin secretion.[38] On the other hand the normal development of hexokinase D can be delayed by various treatments involving the absence of glucose or the existence of diabetes, which will prevent the increase of insulin in the plasma after weaning. Thus induction could be delayed by weaning rats to carbohydrate-free, high-fat and high-protein diets,[39] or by starvation or alloxan-induced diabetes.[33] In addition there have been reports suggesting that glucocorticoids[34] and thyroid hormones[38] could be necessary for precocious hexokinase D induction, but this has not been confirmed.

Thus the ambiguity concerning the role of insulin, thyroid hormones, glucocorticoids and glucose in the regulation of hepatic hexokinase D in the adult rat appears also in the regulation of hepatic hexokinase D development.

Molecular mechanisms involved in hepatic hexokinase D gene expression at weaning

Girard's group have made very interesting studies of the molecular mechanism involved in the regulation of hexokinase D gene expression at weaning,[4] determining the precise time course of the increase of hexokinase D mRNA that occurs after weaning to a high-carbohydrate diet. To avoid the problem of the suckling rats nibbling the mother's food, the mothers were fed a high-fat diet. In these conditions the mRNA does not appear in the liver of the pups until after weaning at 21 days. It increased markedly and reached its maximal value in the first 24 h after weaning to a high-carbohydrate diet.[40] Enzyme activity followed a similar pattern to that of mRNA, but with a lag period of about 24 h, in agreement with Spence's previous results,[10] suggesting that transcriptional regulation plays an important role in setting activity levels.

Weaning produces a rise of insulin concentration in the plasma and a fall of glucagon. When the high-carbohydrate diet was replaced by a high-fat diet, the increase in hexokinase D mRNA activity was reduced by 50%, but not prevented.[40] When the 21-day-old suckling rats were force-fed at weaning with a mixture of sucrose and starch, both glucose and insulin concentrations rapidly increased, reaching maximal values 1h after carbohydrate ingestion, whereas plasma glucagon decreased. Hexokinase D mRNA is undetectable in nuclei isolated before force-feeding of carbohydrate,[40] and begins to accumulate from 2 h afterwards. This change in gene expression and mRNA concentration precedes the changes in enzyme activity, which suggests that the factors responsible for the appearance of liver hexokinase D after weaning act by turning on the transcription of the gene.

On the basis of these data it is reasonable to suppose, in agreement with Girard and colleagues,[4] that the accumulation of hexokinase D mRNA in the liver of rats weaned to a high-carbohydrate diet is due to the increase in insulin secretion. There are three reasons for this:

(i) Physiological levels of insulin stimulate the accumulation of hexokinase D mRNA in primary cultures of suckling rat hepatocytes;

(ii) The lack of an effect of the glucose concentration in the culture media on the accumulation of hexokinase D mRNA in response to insulin (as in adult cells), goes against a direct role of glucose in the accumulation of mRNA;

(iii) Plasma triiodothyronine increases during suckling a long time before hexokinase D mRNA accumulates in the liver, which suggests that this hormone cannot have a direct role in the accumulation though it could contribute to it by potentiating the effects of insulin.

Induction of hexokinase D in neonatal hepatocyte culture

When equivalent studies were done with hepatocytes isolated from 10-12-day-old rats, hexokinase D mRNA was undetectable in such cells after 48 h of culture in serum-free medium devoid of hormones. Addition of insulin or triiodothyronine to the medium results in induction of hexokinase D mRNA.[41] Dexamethasone, on the other hand, does not induce hexokinase D mRNA, though it enhances the response to insulin. The induction by insulin does not depend on the presence of glucose,[41] in agreement with the results with adult hepatocytes described above.[19] The permissive effect of glucocorticoids on the accumulation of hexokinase D mRNA induced by insulin could be due to the stabilizing effects of glucocorticoids on hexokinase D mRNA.

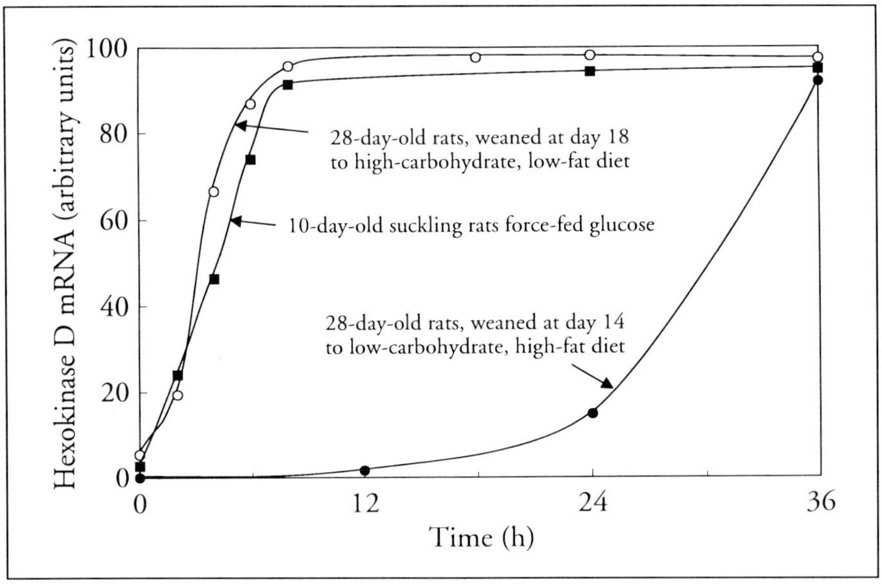

Fig. 8.4. Premature and delayed appearance of hexokinase D mRNA. The effect is shown of insulin on the accumulation of hexokinase D mRNA in cultured hepatocytes from 10-day-old suckling rats that had received an oral glucose load in vivo (■), or from 28-day-old rats weaned onto a high-carbohydrate (○) or a high-fat diet (●). Reproduced with permission from Eur J Biochem 1994; 223:371-380.

There is, however, a difference between the responses to insulin of adult hepatocytes and ones from 10-day-old animals. In cultured adult rat hepatocytes the induction by insulin reaches a maximum at about 6 h, whereas in the cultured 10-day-old animals insulin can only induce hexokinase D mRNA accumulation after a delay of 18 h (see the curve without inhibitors in Fig. 8.5, below).[41] This delay in the response to insulin derives from a delay in transcriptional activation of hexokinase D gene expression as demonstrated by run-on assay, and not from a difference in hexokinase D mRNA half-life,[42] and it is not due to persisting high levels of cAMP during the initial culture period, as it is very low. It can be suppressed if the hepatocytes are obtained from 10-day-old suckling rats that have received an oral administration of glucose (Fig 8.4).[42] The same group previously showed that oral administration of glucose to 10-day-old suckling rats provoked a rapid accumulation of hexokinase D mRNA (within 6 h) by increasing the rate of gene transcription,[43] and that after 18 h both the mRNA and gene transcription had returned to very low values. Consequently in the experiments of Figure 8.4 the hepatocytes were isolated 18 h after the oral administration of glucose. No hexokinase D mRNA was detectable in the hepatocytes at the time of isolation or after plating. Insulin induces a rapid accumulation, with kinetics similar to that observed in adult rat hepatocytes, of hexokinase D mRNA,[19] or in hepatocytes from 10-day-old suckling rats after a secondary exposure to insulin.[42] A similar rapid response is obtained with hepatocytes isolated from 28-day-old rats that have been weaned to a high-carbohydrate diet at day 18; in contrast, if the suckling rats are weaned to a low-carbohydrate diet the response to insulin shows a lag of about 24 h (Fig 8.4).

All these experiments suggest that once hexokinase D has been actively transcribed in vivo, the gene retains the capacity to be rapidly activated by a first exposure to insulin in culture. The mechanism involved is not known, but it could involve the permanent binding of a protein to the transcription initiation complex, this protein being necessary for later activation of hexokinase D gene transcription in response to insulin.[42]

Inhibitors of protein synthesis prevent accumulation of hexokinase D mRNA in response to insulin

If the delay in the initial expression of the hexokinase D gene in response to a primary exposure to insulin is due to the requirement of protein synthesis, this could explain the fact that inhibitors of protein synthesis impair the accumulation. Thus puromycin and cycloheximide, which act at different levels, increase the lag in the response of cultured hepatocytes from 10-day-old suckling rats (Fig 8.5), but do not impair the rapid accumulation of hexokinase D mRNA in response to insulin, in hepatocytes from suckling rats initially cultured with insulin, in hepatocytes from suckling rats given an oral load of glucose in vivo, and in hepatocytes from rats weaned on to a high-carbohydrate diet.[42] Thus the delay in hexokinase D mRNA accumulation in response to primary exposure to insulin could be due to the time required for the synthesis and accumulation of one or several unidentified insulin-dependent proteins necessary for transcription of the hexokinase D gene.

An intriguing point is that hexokinase D is absent in fetal rats, as mentioned above, and furthermore mRNA is undetectable.[4] This deserves a comment, because as mentioned at the beginning of this section the fetus has a high-carbohydrate diet and accordingly a very high insulin concentration in the plasma.[44] However, in spite of the presence of glucose and insulin there is no hexokinase D gene expression. Why? A possible interpretation is that the hexokinase D gene is not in a configuration accessible to transcription factors generated by insulin, because for example of extensive methylation. After birth the regulatory region would become progressively accessible (see below).[4]

EFFECT OF OTHER HORMONES OR SIGNALS

Triiodothyronine

The thyroid hormone triiodothyronine also stimulates transcription of the hexokinase D gene in rats, but its effect is more permis-

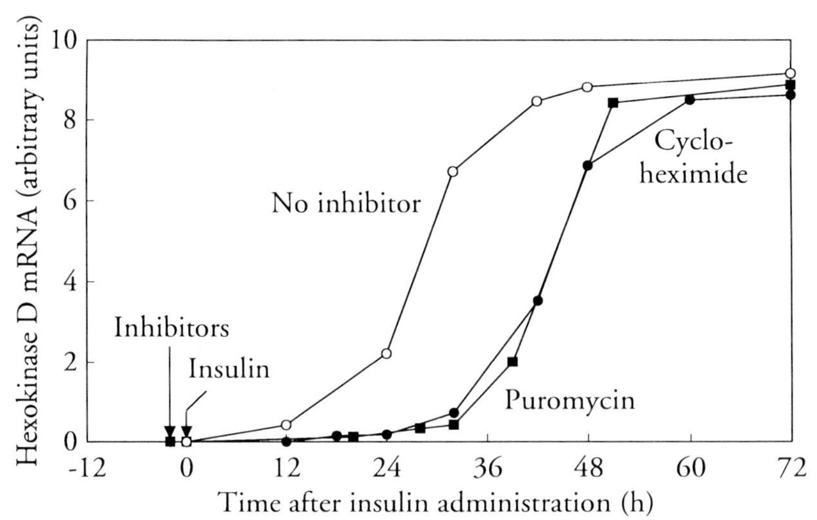

Fig. 8.5. Effect of inhibitors of protein synthesis on the insulin-induced development of hexokinase D mRNA. When puromycin (■) or cycloheximide (●) was added 2 h before the addition of insulin to cultured hepatocytes from 10-day-old suckling rats, the appearance of hexokinase D mRNA was delayed by about 15 h with respect to hepatocytes without inhibitor (○). Reproduced with permission from Eur J Biochem 1994; 223:371-380.

sive. In the whole animal it appears necessary for the induction of hepatic hexokinase D mRNA during the fasting-refeeding transition,[45] but like insulin it has no effect by itself in a starved animal. Thus in hypothyroid rats the response to refeeding was significantly delayed and markedly decreased, whereas the basal levels of hexokinase D mRNA in starved rats were the same in euthyroid and hypothyroid rats. Administration of high doses (50 μg per 100 g body weight) of triiodothyronine restored the levels of hexokinase D mRNA in refed hypothyroid rats to those of euthyroid controls.[45] However, in cultured hepatocytes from newborn rats triiodothyronine has a more striking effect as it is an effective inducer of hexokinase D mRNA,[41] consistent with the precocious induction in neonatal rats previously reported.[38] Thus addition of triiodothyronine to the medium of primary cultures of hepatocytes from 10-12-day-old suckling rats resulted in induction of hexokinase D mRNA, at a level comparable with that elicited by insulin and independent of glucose. The effects of both hormones together were additive. The mechanism underlying the effect of triiodothyronine on hexokinase D mRNA in neonatal hepatocytes has not been determined, but the promotion of de novo appearance of the hexokinase D mRNA suggested an effect at the level of transcription. A direct proof of this point remains to be provided by run-on assays. In any case, the thyroid hormones are anyway known to be involved in the transcriptional regulation of a number of other genes encoding key enzymes of carbohydrate and lipid metabolism.

Dexamethasone

The synthetic glucocorticoid dexamethasone appears to have no effect by itself. However, it enhances the response to insulin, though not that to triiodothyronine.[41]

Biotin

As we saw in Chapter 1, hexokinase D activity is altered by the biotin status of the animal, being low in biotin-deficient rats but increased by biotin administration. This water-soluble vitamin appears to induce transcription of the hepatic hexokinase D gene in rats. A 20-fold increase in the amount of mRNA was observed 1 h after administration of biotin to starved rats, though the effect decayed rapidly and was barely detectable after a further 3 h.[46] Nuclear run-on assays done by the same authors indicated that the transcriptional activity of the gene was increased about 7-fold within 45 min of biotin administration, suggesting that biotin acts at the transcription step. However, this effect may well be indirect and reflect a stimulation of insulin secretion.[1]

PERSPECTIVES

We have seen that in adult rat liver the activity of hexokinase D fluctuates with the nutritional and hormonal status of the animal, and that many factors contribute directly or indirectly to the steady-state activity. Studies in vivo and in vitro have shown that insulin and (via cAMP) glucagon play a crucial role in the regulation of hexokinase D gene transcription, though the precise mechanism remains imperfectly known. It is quite clear that transcription of the gene in hepatocytes depends absolutely on the presence of insulin, at least in the rat, though as we have seen in Chapter 1 we cannot be certain that what is found in the laboratory rat necessarily applies to other species, not even to other rodents. A striking point is that the effect of insulin on transcription is very rapid, as is the repression by cAMP. On the other hand the enzyme is fairly stable in the cell. Does the presence of the regulatory protein to which hexokinase D binds in the absence of glucose contribute to its stability? The use of transgenic mice with decreased amounts of regulatory protein may help to answer this question.

As the adaptive character of hepatic hexokinase D in response to diet is not a conserved feature throughout the vertebrates, as we saw in Chapter 1, it will be very interesting to study the promoter region in species that do not respond to fasting. Comparative studies in the field of hexokinase gene expression will indeed be welcome.

The absence of established hepatic cell lines that express hexokinase D has hampered studies of the promoter region of the gene for hepatic hexokinase D, and this raises a new question: why is expression of the gene extinguished in hepatoma cell lines? So far as the promoter is concerned, the results reported in rats for DNase-I-hypersensitive sites are very interesting,[43] especially in relation to future work. Thus four liver-specific DNase-I-hypersensitive sites have been mapped on 12 kb of the hexokinase D gene between -8 kb upstream and +4 kb downstream of the cap site. They are already present in the fetus immediately before birth, but the intensities of the two proximal sites located upstream from the cap site increase markedly after birth and are fully developed on day 5 after birth. This suggests that the two sites could be implicated in the regulation by insulin and glucagon of hexokinase D gene expression, as full DNase-I hypersensitivity of the two proximal sites seems necessary for the mature response of the gene to changes in the concentrations of the pancreatic hormones.

These four hypersensitive sites are present in adult liver regardless of the transcriptional activity of the hexokinase D gene, i.e. regardless of the nutritional or hormonal status of the adult rat. One possible explanation is that other DNase-I-hypersensitive sites correlated with transcriptional activity may exist upstream or downstream of the promoter region already studied. It is evident that more studies are needed in this area, especially to determine whether similar sites exist in the promoter region of the β-cell hexokinase D gene.

It will be very interesting to examine these DNase-I-hypersensitive sites in patients with MODY. As already mentioned, some such patients have hexokinase D with low specific activity on account of a structural mutation, but it is possible that there may be other cases in which mutations affect the promoter region, thereby impairing transcription and decreasing the hexokinase D activity.

It is possible that in addition to the regulation of transcription of hexokinase D in hepatocytes there may be regulation of translation, as apparently happens in β-cells. Further studies will be needed to clarify this point, in both hepatocytes and β-cells.

The use of transgenic mice is without doubt a valuable tool for studying the physiological consequences of changes in the properties and levels of activity of hexokinase D. This is an emerging field in which some very interesting work has recently been done to study the effect on insulin secretion of varying the glucose-phosphorylating capacity in β-cells, either by expressing yeast hexokinase in these cells[47] or by directly varying the levels of hexokinase D.[48]

As mentioned several times in this book, the concentration of glucose in the blood of mammals is maintained within narrow limits by the concerted effort of three major tissues. The peripheral tissues brain and muscle are the primary sites of glucose uptake and disposal during the postabsorptive and postprandial states respectively. The role of the liver is unique in that it is a net producer of blood glucose via glycogenolysis and gluconeogenesis in the fasted state and a primary site for storage of glucose as glycogen when it is plentiful. The pancreatic islets control the rates of glucose disposal and endogenous glucose production via the regulated secretion of polypeptide hormones, insulin and glucagon, that exert their effects primarily on the liver and skeletal muscle. Hepatocytes and β-cells contain hexokinase D, which is capable of sensing changes in the concentration of glucose in the blood, and to vary its activity instantaneously, the variation being apparently transmitted to the rest of the pathway. The capacity of the β-cell to change its metabolism in response to changes in the blood glucose concentration is crucial for varying the secretion of insulin. For this, hexokinase D appears to be the glucose sensor, as is strongly supported by the linkage with MODY, the diabetes of the young. Fasting decreases hexokinase D activity in the liver and impairs its uptake of glucose. When Hofmeister coined the term "hunger diabetes" more than a century ago he can hardly have imagined how accurate his term would prove to be, or how clear a link would one day be established between glucose phosphorylation and diabetes.

References

1. Iynedjian PB. Mammalian glucokinase and its gene. Biochem J 1993; 293:1-13.
2. Magnuson MA. Tissue-specific regulation of glucokinase gene expression. J Cell Biol 1992; 48:115-121.
3. Vaulont S, Kahn A. Transcriptional control of metabolic regulation genes

by carbohydrates. FASEB J 1994; 8:28-35.
4. Girard J, Decaux JF, Bossard P. Regulation of the initial expression of hepatic phosphoenolpyruvate carboxykinase and glucokinase genes during development. Diabet Metab 1992; 18:74-80.
5. Schudt C. Hormonal regulation of glucokinase in primary cultures of rat hepatocytes. Eur J Biochem 1979; 98:77-82.
6. Katz NR, Nauck MA, Wilson PT. Induction of glucokinase by insulin under the permissive action of dexamethasone in primary rat hepatic cultures. Biochem Biophys Res Comm 1979; 88:23-29.
7. Nakamura T, Aoyama K, Ichihara A. Precocious induction of glucokinase in primary cultures of postnatal rat hepatocytes. Biochem Biophys Res Comm 1979; 91:513-520.
8. Spence JT, Pitot HC. Hormonal regulation of glucokinase in primary cultures of adult rat hepatocytes. J Biol Chem 1979; 254:12331-12336.
9. Spence JT, Merrill MJ, Pitot HC. Role of insulin, glucose, and cyclic GMP in the regulation of glucokinase in cultured hepatocytes. J Biol Chem 1981; 256:1598-1603.
10. Spence JT. Levels of translatable mRNA coding for rat liver glucokinase. J Biol Chem 1983; 258:9143-9146.
11. Iynedjian PB, Ucla C, Mach B. Molecular cloning of glucokinase cDNA. Developmental and dietary regulation of glucokinase mRNA in rat liver. J Biol Chem 1987; 262:6032-6038.
12. Iynedjian PB, Pilot P-R, Nouspikel T et al. Differential expression and regulation of the glucokinase gene in liver and islets of Langerhans. Proc Natl Acad Sci USA 1989; 86:7838-7842.
13. Burch PT, Trus M, Berner DK et al. Adaptation of glycolytic enzymes. Glucose use and insulin release in rat pancreatic islets during fasting and refeeding. Diabetes 1981; 30:923-928.
14. Liang Y, Najafi H, Matschinsky FM. Glucose regulates glucokinase activity in cultured islets from rat pancreas. J Biol Chem 1990; 265:16863-16866.
15. Koranyi L, Bourey RE, Slentz CA et al. Coordinate reduction of rat pancreatric islet glucokinase and proinsulin mRNA by exercise training. Diabetes 1991; 40:401-404.
16. Niemeyer H, Perez N, Codoceo R. Liver glucokinase induction in acute and chronic insulin insufficiency in rats. J Biol Chem 1967; 242:860-864.
17. Iynedjian PB, Gjinovci A, Renold AE. Stimulation by insulin of glucokinase gene transcription in liver of diabetic rats. J Biol Chem 1988; 263:740-744.
18. Reyes A, Rabajille E, Cárdenas ML et al. Stability of hexokinases A, B, and C, and *N*-acetylglucosamine kinase in liver cells isolated from rats submitted to diabetes and several dietary conditions. Biochem J 1984; 221:311-315.
19. Iynedjian PB, Jotterand D, Nouspikel T et al. Transcriptional induction of glucokinase gene by insulin in cultured liver cells and its repression by

glucagon-cAMP system. J Biol Chem 1989; 264:21824-21829.
20. Matsuda T, Noguchi T, Yamada K et al. Regulation of the gene expression of glucokinase and L-type pyruvate kinase in primary cultures of rat hepatocytes by hormones and carbohydrates. J Biochem 1990; 108:778-784.
21. Decaux JF Antoine B, Girard J et al. Regulation of the expression of the L-type pyruvate kinase gene in adult hepatocytes in primary culture. J Biol Chem 1989; 264:11584-11590.
22. Thompson KS, Towle HC. Localization of the carbohydrate response element of the rat L-type pyruvate kinase gene. J Biol Chem 1991; 266:8679-8682.
23. Nouspikel T, Iynedjian PB. Insulin signalling and regulation of glucokinase gene expression in cultured hepatocytes. Eur J Biochem 1992; 210:365-373.
24. Mahadevan LC, Willis AC, Barratt MJ. Rapid histone H3 phosphorylation in response to growth factors, phorbol esters, okadaic acid, and protein synthesis inhibitors. Cell 1991; 65:775-783;.
25. Edwards DR, Mahadevan LC. Protein synthesis inhibitors differentially superinduce c-fos and c-jun by three distinct mechanisms: lack of evidence for labile repressors. EMBO J 1992; 11:2415-2424.
26. Nouspikel T, Gjinovci A, Li S et al. Unimpaired effect of insulin on glucokinase gene expression in hepatocytes challenged with amylin. FEBS Lett 1992; 301:115-118.
27. Flawn P, Loten EG. Properties and distribution of cyclic AMP phosphodiesterase from rat liver. Int J Biochem 1990; 22:983-988.
28. Toth B, Bollen M, Stalmans W. Acute regulation of hepatic protein phosphatases by glucagon, insulin and glucose. J Biol Chem 1988; 263:14061-14066.
29. Dent P, Lavoinne A, Nakielny S et al. The molecular mechanism by which insulin stimulates glycogen synthesis in mammalian skeletal muscle. Nature 1990; 348:302-308.
30. Proud CG. Turned on by insulin. Nature 1994; 371:747-748.
31. Pause A, Belsham GJ, Gingras A-C et al. Insulin-dependent stimulation of protein synthesis by phosphorylation of a regulator of 5'-cap function. Nature 1994; 371:762-767.
32. Ballard JF, Oliver IT. Ketohexokinase, isoenzymes of glucokinase and glycogen synthesis from hexoses in neonatal rat liver. Biochem J 1964; 90:261-268.
33. Walker DG, Holland G. The development of hepatic glucokinase in the neonatal rat. Biochem J 1965; 97:845-854.
34. Jamdar SC, Greengard O. Premature formation of glucokinase in developing rat liver. J Biol Chem 1970; 245:2779-2783.
35. Ureta T, Bravo R, Babul J. Rat liver hexokinases during development. Enzyme 1975; 20:334-348.
36. Hayzer DJ, Iynedjian PB. Alternative splicing of glucokinase mRNA in rat liver. Biochem J 1990; 270:261-263.
37. Haney PM, Estrin CR, Caliendo A et al. Precocious induction of hepatic glucokinase and malic enzyme in artificially reared rat pups fed a high-carbohydrate diet. Arch Biochem Biophys 1986; 244:781-794.

38. Partridge NC, Hoh CH, Weaver PK et al. Premature induction of glucokinase in the neonatal rat by thyroid hormone. Eur J Biochem 1975; 51:49-54.
39. Walker DG, Eaton SW. Regulation of development of hepatic glucokinase in the neonatal rat by the diet. Biochem J 1967; 105:771-777.
40. Perdereau D, Narkewicz MR, Coupe C et al. Hormonal regulation of specific gene expression during development in the rat. Adv Enz Reg 1990; 30:91-108.
41. Narkewicz MR, Iynedjian PB, Ferré P et al. Insulin and tri-iodothyronine induce glucokinase mRNA in primary cultures of neonatal rat hepatocytes. Biochem J 1990; 271:585-589.
42. Bossard P, Decaux J-F, Juanes M et al. Initial expression of glucokinase gene in cultured hepatocytes from suckling rats is linked to the synthesis of an insulin-dependent protein. Eur J Biochem 1994; 223:371-380.
43. Bossard P, Parsa R, Decaux J-F et al. Glucose administration induces the premature expression of liver glucokinase gene in newborn rats. Relation with DNase I hypersensitive sites. Eur J Biochem 1993; 215:883-892.
44. Girard J, Cuendet GS, Marliss EB et al. Fuels, hormones and liver metabolism at term and during the early neonatal period in the rat. J Clin Invest 1973; 52:3190-3200.
45. Höppner W, Seitz H-J. Effect of thyroid hormones on glucokinase gene transcription in rat liver. J Biol Chem 1989; 264:20643-20647.
46. Chauhan J, Dakshinamurti K. Transcription regulation of the glucokinase gene by biotin in starved rats. J Biol Chem 1991; 266:10035-10038.
47. Epstein PN, Boschero AC, Atwater L et al. Expression of yeast hexokinase in pancreatic β cells of transgenic mice reduces blood glucose, enhances insulin secretion, and decreases diabetes. Proc Natl Acad Sci USA 1992; 89:12038-12042.
48. Efrat S, Leiser M, Wu Y-J et al. Ribozyme-mediated attenuation of pancreatic β-cell glucokinase expression in transgenic mice results in impaired glucose-induced insulin secretion. Proc Natl Acad Sci USA 1994; 91:2051-2055.

INDEX

Page numbers in italics denote figures (f) or tables (t).

Molecular Biology Intelligence Unit

Available and Upcoming Titles

- ❒ Organellar Proton-ATPases
  Nathan Nelson, Roche Institute of Molecular Biology
- ❒ Interleukin-10
  Jan DeVries and René de Waal Malefyt, DNAX
- ❒ Collagen Gene Regulation in the Myocardium
  M. Eghbali-Webb, Yale University
- ❒ DNA and Nucleoprotein Structure In Vivo
  Hanspeter Saluz and Karin Wiebauer, HK Institut-Jena and GenZentrum-Martinsried/Munich
- ❒ G Protein-Coupled Receptors
  Tiina Iismaa, Trevor Biden, John Shine, Garvan Institute-Sydney
- ❒ Viroceptors, Virokines and Related Immune Modulators Encoded by DNA Viruses
  Grant McFadden, University of Alberta
- ❒ Bispecific Antibodies
  Michael W. Fanger, Dartmouth Medical School
- ❒ Drosophila Retrotransposons
  Irina Arkhipova, Harvard University and Nataliya V. Lyubomirskaya, Engelhardt Institute of Molecular Biology-Moscow
- ❒ The Molecular Clock in Mammals
  Simon Easteal, Chris Collet, David Betty, Australian National University and CSIRO Division of Wildlife and Ecology
- ❒ Wound Repair, Regeneration and Artificial Tissues
  David L. Stocum, Indiana University-Purdue University
- ❒ Pre-mRNA Processing
  Angus I. Lamond, European Molecular Biology Laboratory
- ❒ Intermediate Filament Structure
  David A.D. Parry and Peter M. Steinert, Massey University-New Zealand and National Institutes of Health
- ❒ Fetuin
  K.M. Dziegielewska and W.M. Brown, University of Tasmania
- ❒ Drosophila Genome Map: A Practical Guide
  Daniel Hartl and Elena R. Lozovskaya, Harvard University
- ❒ Mammalian Sex Chromosomes and Sex-Determining Genes
  Jennifer A. Marshall-Graves and Andrew Sinclair, La Trobe University-Melbourne and Royal Children's Hospital-Melbourne
- ❒ Regulation of Gene Expression in *E. coli*
  E.C.C. Lin, Harvard University
- ❒ Muscarinic Acetylcholine Receptors
  Jürgen Wess, National Institutes of Health
- ❒ Regulation of Glucokinase in Liver Metabolism
  Maria Luz Cardenas, CNRS-Laboratoire de Chimie Bactérienne-Marseille
- ❒ Transcriptional Regulation of Interferon-γ
  Ganes C. Sen and Richard Ransohoff, Cleveland Clinic
- ❒ Fourier Transform Infrared Spectroscopy and Protein Structure
  P.I. Haris and D. Chapman, Royal Free Hospital-London
- ❒ Bone Formation and Repair: Cellular and Molecular Basis
  Vicki Rosen and R. Scott Thies, Genetics Institute, Inc.-Cambridge
- ❒ Mechanisms of DNA Repair
  Jean-Michel Vos, University of North Carolina
- ❒ Short Interspersed Elements: Complex Potential and Impact on the Host Genome
  Richard J. Maraia, National Institutes of Health
- ❒ Artificial Intelligence for Predicting Secondary Structure of Proteins
  Xiru Zhang, Thinking Machines Corp-Cambridge
- ❒ Growth Hormone, Prolactin and IGF-I as Lymphohemopoietic Cytokines
  Elisabeth Hooghe-Peters and Robert Hooghe, Free University-Brussels
- ❒ Human Hematopoiesis in SCID Mice
  Maria-Grazia Roncarolo, Reiko Namikawa and Bruno Péault DNA Research Institute
- ❒ Membrane Proteases in Tissue Remodeling
  Wen-Tien Chen, Georgetown University
- ❒ Annexins
  Barbara Seaton, Boston University
- ❒ Retrotransposon Gene Therapy
  Clague P. Hodgson, Creighton University
- ❒ Polyamine Metabolism
  Robert Casero Jr, Johns Hopkins University
- ❒ Phosphatases in Cell Metabolism and Signal Transduction
  Michael W. Crowder and John Vincent, Pennsylvania State University
- ❒ Antifreeze Proteins: Properties and Functions
  Boris Rubinsky, University of California-Berkeley
- ❒ Intramolecular Chaperones and Protein Folding
  Ujwal Shinde, UMDNJ
- ❒ Thrombospondin
  Jack Lawler and Jo Adams, Harvard University
- ❒ Structure of Actin and Actin-Binding Proteins
  Andreas Bremer, Duke University
- ❒ Glucocorticoid Receptors in Leukemia Cells
  Bahiru Gametchu, Medical College of Wisconsin
- ❒ Signal Transduction Mechanisms in Cancer
  Hans Grunicke, University of Innsbruck
- ❒ Intracellular Protein Trafficking Defects in Human Disease
  Nelson Yew, Genzyme Corporation
- ❒ apoJ/Clusterin
  Judith A.K. Harmony, University of Cincinnati
- ❒ Phospholipid Transfer Proteins
  Vytas Bankaitis, University of Alabama
- ❒ Localized RNAs
  Howard Lipschitz, California Institute of Technology
- ❒ Modular Exchange Principles in Proteins
  Laszlo Patthy, Institute of Enzymology-Budapest
- ❒ Molecular Biology of Cardiac Development
  Paul Barton, National Heart and Lung Institute-London
- ❒ RANTES, *Alan M. Krensky, Stanford University*
- ❒ New Aspects of V(D)J Recombination
  Stacy Ferguson and Craig Thompson, University of Chicago

MEDICAL INTELLIGENCE UNIT

AVAILABLE AND UPCOMING TITLES

- ❒ Hyperacute Xenograft Rejection
  Jeffrey Platt, Duke University
- ❒ Chimerism and Tolerance
  Suzanne Ildstad, University of Pittsburgh
- ❒ Birth Control Vaccines
  G.P. Talwar and Raj Raghupathy, National Institute of Immunology-New Delhi and University of Kuwait
- ❒ Monoclonal Antibodies in Transplantation
  Lucienne Chatenoud, Hôpital Necker-Paris
- ❒ Therapeutic Applications of Oligonucleotides
  Stanley Crooke, ISIS Pharmaceuticals
- ❒ Cryopreserved Venous Allografts
  Kelvin G.M. Brockbank, CryoLife, Inc.
- ❒ Clinical Benefits of Leukpodepleted Blood Products
  Joseph Sweeney and Andrew Heaton, Miriam and Roger Williams Hospitals-Providence and Irwin Memorial Blood Center-San Francisco
- ❒ Delta Hepatitis Virus
  M. Dinter-Gottlieb, Drexel University
- ❒ Intima Formation in Blood Vessels: Spontaneous and Induced
  Mark M. Kockx, Algemeen Ziekenhuis Middelheim-Antwerpen
- ❒ Adult T Cell Leukemia and Related Diseases
  Takashi Uchiyama and Jungi Yodoi, University of Kyoto
- ❒ Development of Epstein-Barr Virus Vaccines
  Andrew Morgan, University of Bristol
- ❒ p53 Suppressor Gene
  Tapas Mukhopadhyay, Steven Maxwell and Jack A. Roth, University of Texas-MD Anderson Cancer Center
- ❒ Sudden Death in Ischemic Heart Disease
  Malcolm D. Silver and Giorgio Baroldi, University of Toronto and University of Milan
- ❒ Minor Histocompatibility Antigens and Transplantation
  Craig V. Smith, University of Pittsburgh
- ❒ Familial Adenomatous Polyposis Coli and the APC Gene
  Joanna Groden, University of Cincinnati
- ❒ Cancer Cell Adhesion and Tumor Invasion
  Pnina Brodt, McGill University
- ❒ Constitutional Immunity to Infection
  Cees M. Verduin, David A. Watson, Jan Verhoef, Hans Van Dijk, University of Utrecht and North Dakota State University
- ❒ Nutritional and Metabolic Support in Critically Ill Patients
  Rifat Latifi and Stanley Dudrick, Yale University
- ❒ Nutritional Support in Gastrointestinal Failure
  Rifat Latifi and Stanley Dudrick, Yale University
- ❒ Septic Myocardiopathy: Molecular Mechanisms
  Karl Werdan, Ludwig-Maximilians-Universität-München
- ❒ The Molecular Genetics of Wilms Tumor
  Max Coppes, Christine Campbell and Bryan R.G. Williams, Cleveland Clinic and University of Calgary
- ❒ Endothelins
  David J. Webb and Gillian Gray, University of Edinburgh
- ❒ Nutritional and Metabolic Support in Cancer, Transplant and Immunocompromised Patients
  Rifat Latifi, Yale University
- ❒ Antibody-Mediated Graft Rejection
  J. Andrew Bradley and Eleanor Bolton, University of Glasgow
- ❒ Liposomes in Cancer Chemotherapy
  Steven Sugarman, University of Texas-MD Anderson
- ❒ Molecular Basis of Human Hypertension
  Florent Soubrier, Collége de France-Paris
- ❒ Endocardial Endothelium: Control of Cardiac Performance
  Stanislas U. Sys and Dirk Brutsaert, Universiteit Antwerpen
- ❒ Endovascular Stented Grafts for the Treatment of Vascular Diseases
  Michael L. Marin, Frank J. Veith and Barry A. Levine, Albert Einstein College of Medicine
- ❒ B Cells and Autoimmunity
  Christian Boitard, Hôpital Necker-Paris
- ❒ Immunity to Mycobacteria
  Ian Orme, Colorado State University
- ❒ Hepatic Stem Cells and the Origin of Hepatic Carcinoma
  Stewart Sell, University of Texas-Houston
- ❒ HLA and Maternal-Fetal Recognition
  Joan S. Hunt, University of Kansas
- ❒ Transplantation Tolerance
  J. Wesley Alexander, University of Cincinnati
- ❒ Ovarian Autoimmunity
  Roy Moncayo and Helga E. Moncayo, University of Innsbruck
- ❒ Ex Vivo Cell Therapy for Cancer
  Ronald Herberman, University of Pittsburgh
- ❒ Protein and Amino Acid Metabolism in Cancer
  Peter W.T. Pisters and Murray Brennan, Sloan-Kettering Memorial Cancer Center
- ❒ Cytokines and Hemorrhagic Shock
  Eric J. DeMaria, Medical College of Virginia
- ❒ Endometriosis
  Lisa Barrie Schwartz, New York University and David Olive, Yale University
- ❒ T Cell Vaccination and Autoimmune Disease
  Jingwu Zhang, Willems Institut-Belgium
- ❒ Immune Privilege
  J. Wayne Streilein, Luke Jiang and Bruce Ksander, Schepens Eye Research Institute-Boston
- ❒ The Pathophysiology of Sepsis and Multi-Organ Failure
  Mitchell Fink, Harvard University
- ❒ Bone Metastasis
  F. William Orr, McMaster University
- ❒ Novel Regional Therapies for Liver Tumors
  Seiji Kawasaki and Masatoshi Makuuchi, Shinshu University
- ❒ Thyroid Hormone Resistance
  Roy E. Weiss and Michael Refetoff, University of Chicago
- ❒ Growth Hormone in Critical Illness
  Michael Torosian, University of Pennsylvania
- ❒ Molecular Biology of Aneurysms
  Richard R. Keen, Northwestern University
- ❒ Strategies in Malaria Vaccine Design
  F.E.G. Cox, King's College London
- ❒ Chimeric Proteins and Fibrinolysis
  Christoph Bode, Marschall Runge and Edgar Haber, University of Heidelberg, University of Texas-Galveston and Harvard University

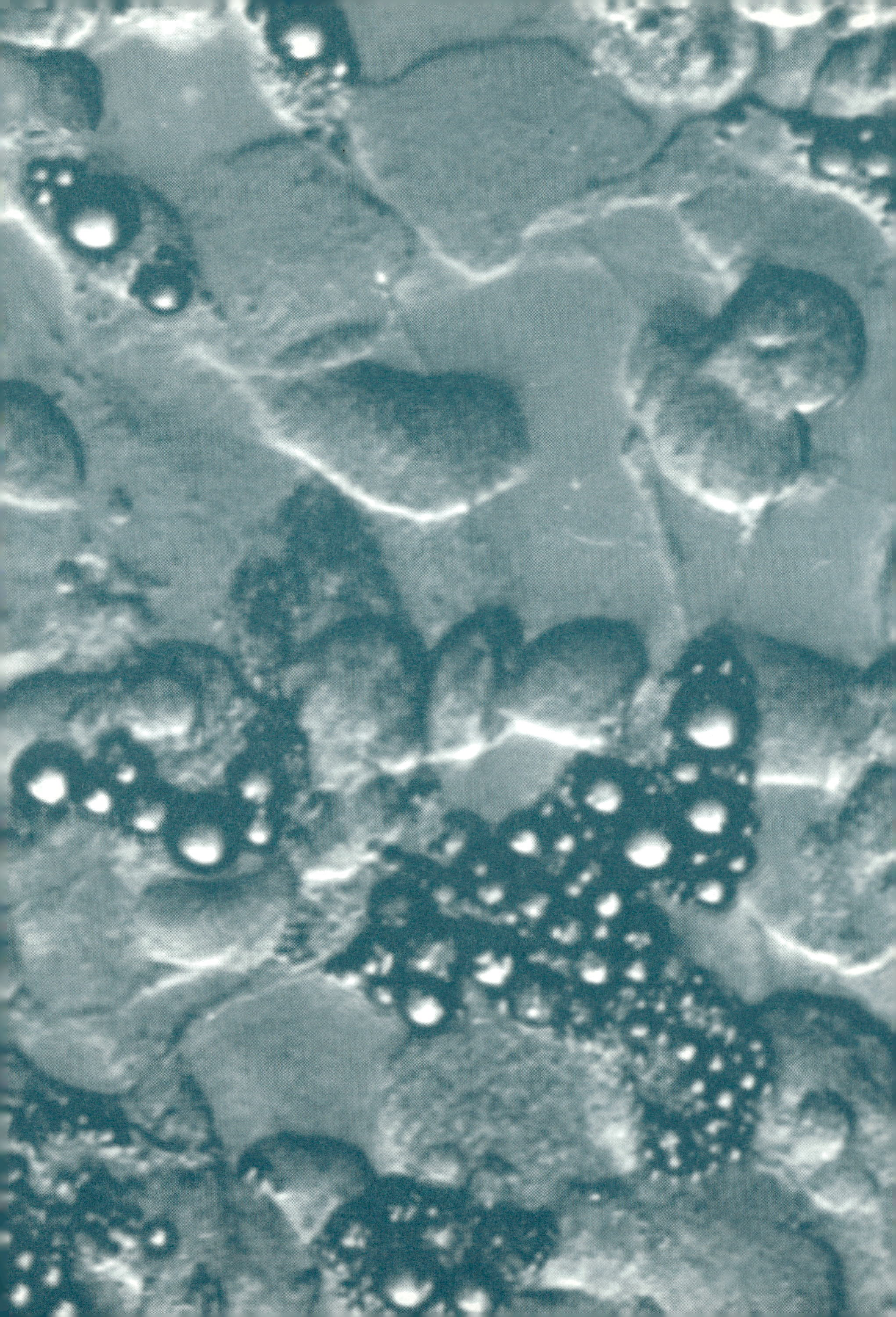